Lecture Notes in Statistics 203

Edited by P. Bickel, P.J. Diggle, S.E Fienberg, U. Gather,
I. Olkin, S. Zeger

T0223395

Pierre Alquier • Eric Gautier • Gilles Stoltz
Editors

Inverse Problems and High-Dimensional Estimation

Stats in the Château Summer School,
August 31 - September 4, 2009

Springer

Editors
Pierre Alquier
Université Paris 7
Laboratoire de Probabilités et
Modèles Aléatoires
175 rue du Chevaleret
75205 Paris Cedex 13
France
alquier@math.jussieu.fr

Eric Gautier
ENSAE
Centre de Recherche en Economie et Statistique
3 avenue Pierre Larousse
92245 Malakoff
France
eric.gautier@ensae-paristech.fr

Gilles Stoltz
HEC Paris
Department of Economics and Decision Sciences
1 rue de la Libération
78351 Jouy-en-Josas
&
Ecole Normale Supérieure
Département de Mathématiques et Applications
45 rue d'Ulm
75005 Paris
France
stoltz@hec.fr

ISSN 0930-0325
ISBN 978-3-642-19988-2   e-ISBN 978-3-642-19989-9
DOI 10.1007/978-3-642-19989-9
Springer Heidelberg Dordrecht London New York

Library of Congress Control Number: 2011930794

Cover design: eStudio Calamar S.L.

Printed on acid-free paper

Springer is part of Springer Science+Business Media (www.springer.com)

Preface

The "Stats in the Château" Summer School

The "Stats in the Château" summer school was held at the CRC château on the campus of HEC Paris, Jouy-en-Josas, France, from August 31 to September 4, 2009. It was organized jointly by faculty members of three French academic institutions: ENSAE ParisTech, Ecole Polytechnique ParisTech, and HEC Paris. These institutions cooperate through a scientific foundation devoted to the decision sciences.

The summer school brought together about 70 researchers and PhD students in economics, statistics, mathematics and computer science, all interested in both mathematical statistics and applications to economics. The motto was that economics is a source of interesting new problems for statisticians and that, conversely, recent statistical methods, sometimes motivated by other fields, can be used for quantitative analysis in economics. The goal was therefore to introduce the audience both to some modern sets of methods and to a wide range of their applications to economics, and to foster discussions between statisticians and economists.

The scientific content of the summer school focused on two themes: inverse problems and high-dimensional estimation. Two courses were given, one by Laurent Cavalier (Université Aix-Marseille I) on ill-posed inverse problems, and one by Victor Chernozhukov (Massachusetts Institute of Technology) on high-dimensional estimation with applications to economics. Ten invited lecturers – whose names can be found in the appendix of this volume together with the titles of their talk – illustrated the two courses and provided either reviews of the state of the art in the field or of applications and original research contributions. The opportunity was also given to participants to present their own findings either in the form of a contributed talk or during a poster session held on the first day of the summer school.

Inverse Problems and High-Dimensional Estimation

The first theme of the summer school was ill-posed statistical inverse problems. This is already considered an important problem in many areas of science, and also became an important issue in econometrics about ten years ago. Nonparametric specifications are important to allow for flexible models. Statistical inverse problems are problems from nonparametric statistics. A wide class of models from economics can be formulated as inverse problems, that is, as a relation between a function, directly related to the observations, and a structural functional parameter. Examples include the estimation of the distributions of the following objects: types in a game-theoretical model where only actions of the players are observed; random coefficients accounting for unobserved heterogeneity; the pricing operator given observed option prices; a regression function in the presence of endogenous regressors; etc. Because inversion often leads to a lack of continuity, the inference requires some suitable regularization. Theoretical properties such as optimal rates of convergence and adaptation are important to study.

The second theme was high-dimensional estimation. High-dimensionality corresponds to the case where the parameter of interest has a dimension p possibly much larger than the sample size n. A lot of attention has been given to this setting in recent years in the statistics and machine learning communities. In this setting, parsimonious models can still be estimated. Parsimony is also referred to as sparsity and corresponds to the case where, though the number of parameters is very large, only a small number of them are non-zero. Results can often be extended to the case where most parameters are too small to matter. This is a setting often encountered in the social sciences. Several techniques have been developed to extract relevant parameters from large vectors, in particular, the Lasso, the Dantzig selector, and Bayesian-type methods. These techniques have been widely implemented in imaging and bioinformatics. At the time of the summer school the use of the above-mentioned methods in economics, while widely applicable, was very limited.

The Proceedings

After the summer school, the twelve researchers giving the lectures or the invited talks were given the opportunity to contribute to the present volume. The aim was to provide an accessible but rigorous mathematical introduction to these two modern sets of problems from statistics and econometrics, and to present applications to quantitative problems in economics. The intended audience is the same as that of the summer school: young researchers, e.g., PhD students in statistics and economics, or more senior researchers from related fields.

The book brings together contributions from five invited speakers with their coauthors, among them, the two lecturers. Laurent Cavalier provides detailed lecture notes on ill-posed statistical inverse problems while Victor Chernozhukov reviews Lasso-based methods for estimating high-dimensional regression models with applications to empirical economic problems. These lecture notes are illustrated and further developed by three other contributions. On the one hand, Jean-Pierre Florens discusses the case of nonparametric estimation with endogenous variables using instrumental variables. On the other hand, Felix Abramovich and Ya'acov Ritov respectively present a model selection and a Bayesian viewpoint on high-dimensional estimation.

Interested readers can find the slides of most of the invited and contributed talks, as well as the videotape of the first lecture by Laurent Cavalier, on the website of the summer school http://www.hec.fr/statsinthechateau.

Acknowledgments

The scientific committee of the summer school included

Christian Gourieroux	(ENSAE-CREST – Université de Toronto)
Yuichi Kitamura	(Yale University)
Alexandre Tsybakov	(ENSAE-CREST)

The summer school was mainly funded by the "Groupement d'Intérêt Scientifique: Sciences de la Décision", a scientific foundation devoted to the decision sciences, gathering Ecole Polytechnique ParisTech, ENSAE ParisTech and HEC Paris.

It was also supported by the EDF-Calyon "Finance et développement durable" (finance and sustainable development) chair.

The faculty members in charge of the local organization were

Pierre Alquier	(ENSAE-CREST – Université Paris Diderot)
Veronika Czellar	(HEC Paris)
Alfred Galichon	(Ecole Polytechnique ParisTech)
Eric Gautier	(ENSAE-CREST)
Gilles Stoltz	(CNRS – Ecole normale supérieure – HEC Paris)

and they were assisted by

Nathalie Beauchamp	(HEC Paris)
Claudine Tantillo	(HEC Paris)

Paris, January 2011

Pierre Alquier
Eric Gautier
Gilles Stoltz

Contents

List of Contributors

Felix Abramovich
Tel Aviv University, Department of Statistics & Operations Research, Ramat Aviv, Tel Aviv 69978, Israel, e-mail: felix@post.tau.ac.il

Alexandre Belloni
Duke University, Fuqua School of Business, 100 Fuqua Drive, Durham, NC 27708, USA, e-mail: abn5@duke.edu

Natalia Bochkina
University of Edinburgh, School of Mathematics, King's Buildings, Mayfield Road, Edinburgh, EH9 3JZ, UK, e-mail: N.Bochkina@ed.ac.uk

Laurent Cavalier
Université Aix-Marseille 1, LATP, CMI, 39 rue Joliot-Curie, 13453 Marseille, France, e-mail: cavalier@cmi.univ-mrs.fr

Victor Chernozhukov
Massachusetts Institute of Technology, Department of Economics, 50 Memorial Drive, Cambridge, MA 02142, USA, e-mail: vchern@mit.edu

Jean-Pierre Florens
Université Toulouse 1 & Toulouse School of Economics, GREMAQ & IDEI, 21 allée de Brienne, 31000 Toulouse, France, e-mail: florens@cict.fr

Vadim Grinshtein
The Open University of Israel, Department of Mathematics, P.O.Box 808, Raanana 43107, Israel, e-mail: vadimg@openu.ac.il

Ya'acov Ritov
The Hebrew University of Jerusalem, Department of Statistics, Mount Scopus, Jerusalem 91905, Israel, e-mail: yaacov.ritov@gmail.com

Part I
Lecture Notes on Inverse Problems

Chapter 1
Inverse Problems in Statistics

Laurent Cavalier

Abstract There exist many fields where inverse problems appear. Some examples
are: astronomy (blurred images of the Hubble satellite), econometrics (instrumental variables), financial mathematics (model calibration of the volatility), medical
image processing (X-ray tomography), and quantum physics (quantum homodyne
tomography).

These are problems where we have indirect observations of an object (a function)
that we want to reconstruct, through a linear operator A. Due to its indirect nature,
solving an inverse problem is usually rather difficult.

For this reason, one needs regularization methods in order to get a stable and
accurate reconstruction.

We present the framework of statistical inverse problems where the data are cor-
rupted by some stochastic error. This white noise model may be discretized in the
spectral domain using Singular Value Decomposition (SVD), when the operator A
is compact. Several examples of inverse problems where the SVD is known are
presented (circular deconvolution, heat equation, tomography).

We explain some basic issues regarding nonparametric statistics applied to inverse problems. Standard regularization methods and their counterpart as estimation procedures by use of SVD are discussed (projection, Landweber, Tikhonov,
...). Several classical statistical approaches like minimax risk and optimal rates of
convergence, are presented. This notion of optimality leads to some optimal choice
of the tuning parameter.

However these optimal parameters are unachievable since they depend on the
unknown smoothness of the function. This leads to more recent concepts like adaptive estimation and oracle inequalities. Several data-driven selection procedures of
the regularization parameter are discussed in details, among these: model selection
methods, Stein's unbiased risk estimation and the recent risk hull method.

Laurent Cavalier
Université Aix-Marseille 1, LATP, CMI, 39 rue Joliot-Curie, 13453 Marseille, France, e-mail:
cavalier@cmi.univ-mrs.fr

Preface

These notes are based on a mini-course which was given during the summer school *Stats in the Château* in August 2009. The first version of these notes was written for a course at Heidelberg University in 2007. Another course was given at *Ecole d' été en statistique* in Switzerland in September 2010. A longer version of the course is given to the graduate students at Université de Provence in Marseille.

I would like to thank the colleagues and students who attended these courses and asked questions, made comments and remarks.

Since these notes were written in several places, I would also like to thank Heidelberg University, Göttingen University, Sydney University, University College London and Princeton University.

Many thanks, for very helpful discussions, to Yuri Golubev, Markus Reiss and a special thank to Thorsten Hohage for giving to me his lecture notes.

The three referees also helped a lot, with their very interesting remarks and comments, in improving these notes.

I would like to dedicate these notes to two absentees:

To Marc Raimondo, I will not join you any more in Sydney to write a book on inverse problems and wavelets;

To my father, I know you would have loved...

Marseille, January 2011   Laurent Cavalier

1.1 Inverse Problems

1.1.1 Introduction

There exist many fields of sciences where inverse problems appear. Some examples are: astronomy (blurred images of the Hubble satellite), econometrics (instrumental variables), financial mathematics (model calibration of the volatility), medical image processing (X-ray tomography), and quantum physics (quantum homodyne tomography)

These are problems where we have indirect observations of an object (a function) that we want to reconstruct. The common structure of all these problems, coming from very different fields, is that we only have access to indirect observations. Due to its indirect nature, solving an inverse problem is usually rather difficult. In fact, there is a need for accurate methods, called regularization methods, in order to solve such an inverse problem.

One example is the problem of X-ray tomography (see Section 1.1.6.5). In this framework, the goal is to reconstruct the internal structure of a human body, by use of external observations. Thus, the internal image cannot be observed directly, but only indirectly.

This notion of indirect observations of some function is usually modeled by use of an operator A. From a mathematical point of view, inverse problems usually correspond to the inversion of this operator.

Let A be a bounded operator from H into G, where H and G are two separable Hilbert spaces. The classical problem is the following.

$$\text{Given } g \in G, \text{ find } f \in H \text{ such that } Af = g. \qquad (1.1)$$

The terminology of inverse problem comes from the fact that one has to invert the operator A. A case of major interest is the case of ill-posed problems where the operator is not invertible. The issue is then to handle this inversion in order to obtain a precise reconstruction.

A classical definition is the following (see [65]).

Definition 1.1. A problem is called **well-posed** if

1. there exists a solution to the problem (existence);
2. there is at most one solution to the problem (uniqueness);
3. the solution depends continuously on the data (stability);

A problem which is not well-posed is called **ill-posed**.

One is usually not too much concerned with the existence. If the data space is defined as the set of solutions, existence is clear. Otherwise, the concept of solution may be slightly changed.

If uniqueness is not verified, this is more serious. If there exist several solutions then one has to decide which one is of interest or give additional information. However, the problem of uniqueness is usually relevant in inverse problems.

A standard way of solving the existence and uniqueness problems is by resorting to generalized inverses (see Section 1.1.4).

These two problems (existence and uniqueness) are similar to the standard problem of identifiability in statistics.

Nevertheless, the main issue is usually stability. Indeed, suppose A^{-1} exists but is not bounded. Given a noisy version of g called g_ε, the reconstruction $f_c = A^{-1} g_\varepsilon$ may be far from the true f. Thus, one needs to invert the operator A in a more stable way. Therefore, one has to develop regularization methods, in order to get fine reconstructions even in ill-posed problems.

A century ago it was generally believed that for natural problems the solution would always depend continuously on the data. Otherwise the mathematical model was believed to be inadequate. These problems therefore were called ill-posed. The idea was that the problem was genuiely not well-posed and that there was no chance to solve such a problem. Ill-posed problems were usually considered, more or less, as unsolvable problems.

Only sixty years ago, scientists realized that a large number of problems which appeared in sciences were ill-posed in any framework. The idea was developed that there was natural ill-posed problems, in the sense that these were ill-posed in any setting, but they could be however solved by use of regularization methods.

This initiated a lot of research in order to get accurate regularization methods, see for example [127, 123, 128, 108, 10, 49, 110, 117, 116, 126, 51, 72, 112].

1.1.2 Statistical Inverse Problems

Loosely speaking solving an inverse problem means recovering an object f from indirect noisy observations Y. The object f is usually modeled as a function (or a vector) that has been modified by an operator A; thus one observes a noisy version of Af. From a mathematical point of view, solving the inverse problem boils down to inverting the operator A. The problem is that A may not be invertible or nearly so. This is the case of ill-posed problems and it is of great practical interest as it arises naturally in many fields such as geophysics, finance, astronomy, biology, . . .

Ill-posed problems are further compounded by the presence of errors (noise) in the data. Statistics enters inverse problems when at least one of the components of the inverse problem (usually the noise) is modeled as stochastic. The question is then to study statistical regularization methods that lead to a meaningful reconstruction despite the noise and ill-posedness.

In Section 1.1 we will present the standard framework of inverse problems focusing on linear operator and stochastic noise. Basic notions on operator theory will be recalled, especially the case of compact operators and singular value decomposition. However, the spectral theory and functional calculus will be defined even for non-compact operators. Several examples of standard inverse problems will be given.

In our opinion the inverse problem framework is better known among statisticians than its statistical approach is among the inverse problem community. For instance, the latter is well acquainted with the concepts of mean, variance and bias but is less familiar with classical concepts such as white noise model, risk estimation, minimax risk, model selection and optimal rates of convergence, which we will discuss in Section 1.2. In addition to these classical notions we will present in Section 1.3 some more recent concepts that have been developed since the 90s like adaptive estimation, oracle inequalities, model selection methods, Stein's unbiased risk estimation and the recent risk hull method. Section 1.4 is a conclusion. We will discuss on the topics which we think are important in the statistical study of inverse problems. Moreover, several open problems will be presented in order to go beyond the framework of these lectures.

All the statistical concepts will be defined and discussed in the framework of inverse problems. Although some of the techniques are specific to this field, some may also be used in more general situations. Other statistical methods not discussed in these notes may also have applications to inverse problems but one should be careful with their application given the intrinsic difficulty and instability of ill-posed problems.

In our mind this is one of the most appealing points of statistical inverse problems. Indeed, most of the standard problems in nonparametric statistics are present in this framework. One may study estimation methods, minimax estimation, rates of convergence for different functional classes (Besov balls, Hölder balls, Sobolev balls), various risk assessments (L^2, L^p, pointwise risk). One may also study more recent notions like adaptive estimation, model selection, data-driven selection methods, oracle inequalities, and so on.

On the other hand, there exist also many problems which are specific to the framework of inverse problems. One can consider, noise in the operator, or the problem of choosing the best basis for a given operator. Moreover, due to the ill-posedness and the difficulty of inverse problems, building accurate estimators is usually much more involved here than in the direct problem.

The aim of these notes is to explain some standard theoretical issues regarding the statistical framework of inverse problems. These lectures provide a glimpse of modern nonparametric statistics in the context of inverse problems. Other topics and reviews may be found in [108, 110, 116, 126, 51, 81, 23].

1.1.3 Linear Inverse Problems with Random Noise

The classical framework for inverse problem is given by linear inverse problems between two Hilbert spaces.

Let H and G two separable Hilbert spaces. Let A be a known linear bounded operator from the space H to G.

Suppose that we have the following observation model

$$Y = Af + \varepsilon \xi, \tag{1.2}$$

where Y is the observation, f is an unknown element in H, ξ is an error, ε corresponds to the noise level. Our aim here is to estimate (or reconstruct) the unknown f by use of the observation Y. The idea is that, at least when ε is small, rather sharp reconstruction should be obtained.

The standard framework first considered by [127] and further studied by [128] corresponds to the case of inverse problems with deterministic noise. In this case, the noise ξ is considered as some element in G, with $\|\xi\| \leqslant 1$. Since the noise is some unknown element of a ball in G, the results have to be obtained for any possible noise, i.e. for the worst noise. The study of deterministic noise is not the aim of these notes and may be found in Section 1.2.5.

Our framework is a statistical inverse problem, which was considered in [123]. Indeed we observe a noisy version (with random error) of Af and we want to reconstruct f. Thus, three main difficulties appear:

- dealing with the noise in the observation (statistics);
- inverting the operator A (inverse problems theory);
- deriving numerical implementations (computational mathematics);

Our aim is now to propose reasonable assumptions on the stochastic noise. The stochastic error is a Hilbert-space process, i.e. a bounded linear operator $\xi : G \rightarrow L^2(\Omega, \mathscr{A}, P)$ where $(\Omega, \mathscr{A}, P)$ is the underlying probability space and $L^2(\cdot)$ is the space of all square integrable measurable functions.

Thus, for all functions $g_1, g_2 \in G$, the random variables $\langle \xi, g_j \rangle$ $j = 1, 2$ are defined, by definition $\mathbf{E}\langle \xi, g_j \rangle = 0$ and define its covariance Cov_ξ as the bounded linear operator ($\|\mathrm{Cov}_\xi\| \leqslant 1$) from G in G such that $\langle \mathrm{Cov}_\xi g_1, g_2 \rangle = \mathrm{Cov}(\langle \xi, g_1 \rangle, \langle \xi, g_2 \rangle)$.

A Hilbert-space random variable $\varkappa$ is a measurable function: $\Omega \to G$. Any Hilbert-space random variable with a finite second moment may be identified with an Hilbert-space process by defining $\varphi \to \langle \varkappa, \varphi \rangle$. However, not all Hilbert-space processes are Hilbert-space random variables.

The action of an operator $A \in L(G, H)$ on some Hilbert-space process ξ is given in Definition 1.3.

The standard hypothesis, which will be mainly considered in these notes, corresponds to the following assumption.

Definition 1.2. We say that ξ is a **white noise** process in G, if $\mathrm{Cov}_\xi = I$ and the induced random variables are Gaussian:

for all functions $g_1, g_2 \in G$, the random variables $\langle \xi, g_j \rangle$ have distributions $\mathcal{N}(0, \|g_j\|^2)$ and $\mathrm{Cov}(\langle \xi, g_1 \rangle, \langle \xi, g_2 \rangle) = \langle g_1, g_2 \rangle$.

See for example [69].

The white noise is one of the more standard stochastic noise considered in statistics, see for example the Gaussian white noise model in Section 1.1.6.1.

One of the main property of a white noise process is the following.

Lemma 1.1. *Let ξ be a white noise in G and $\{\psi_k\}$ be an orthonormal basis in G. Define ξ_k by $\xi_k = \langle \xi, \psi_k \rangle$. Then $\{\xi_k\}$ are i.i.d. standard Gaussian random variables.*

Proof. By definition $\xi_k \sim \mathcal{N}(0, \|\psi_k\|^2) = \mathcal{N}(0, 1)$. Moreover, we have $\mathbf{E}(\langle \xi, \psi_k \rangle, \langle \xi, \psi_\ell \rangle) = \langle \psi_k, \psi_\ell \rangle = \delta_{k\ell}$. Note also that $\{\xi_k\}$ is Gaussian.

Remark 1.1. This lemma is very important and almost characterizes a white noise. Indeed, by projection on some orthonormal basis $\{\psi_k\}$, one obtains a sequence of i.i.d. standard Gaussian random variables $\{\xi_k\}$. This is a way to understand the notion of white noise in applications. In a model with white noise, one obtains a standard Gaussian i.i.d. noise in each observed coefficient (see Section 1.1.5).

Remark 1.2. Another remark is that a white noise, as a Hilbert-space process, is not in general a Hilbert-space random variable; note also that $\|\xi\|_G = \infty$, thus ξ is not an element of G. One main difference between the deterministic and the stochastic approaches of inverse problems is that the random noise is large compared to the deterministic one. This discussion is postponed to Section 1.2.5.

Note that when ξ is a white noise, Y does not belong to G, but acts on G, with the following definition, which follows from (1.2),

$$\forall \psi \in G, \quad \langle Y, \psi \rangle = \langle Af, \psi \rangle + \varepsilon \langle \xi, \psi \rangle,$$

where $\langle \xi, \psi \rangle \sim \mathcal{N}(0, \|\psi\|^2)$.

Remark 1.3. White noise may also be identified with a generalized random variable. Indeed, it does not take its values in G but acts on G, see [69].

1.1.4 Basic Notions on Operator Theory

Operator theory contains the basic mathematical tools that are needed in inverse problems. In this section we recall rather quickly some standard notions on operator theory which will be used through these lectures. We concentrate on linear bounded operators between Hilbert spaces.

Let H and G be two separable Hilbert spaces.

Definition 1.3. 1. A is a **bounded (or continuous) linear operator** from H to G if it is a linear application from $D(A) = H$ to G which is continuous on H.
2. Denote by $D(A)$ the definition domain of A, by $R(A) = A(H)$ its range, by $N(A) = \{\varphi \in H : A\varphi = 0\}$ its null-space, by $L(H, G)$ the space of linear bounded operators from H to G and by $\|A\|$ the operator norm $\|A\| = \sup\{\|A\varphi\| : \|\varphi\| = 1\}$.
3. The operator $A \in L(H, G)$ is said to be **invertible** if there exists A^{-1} in $L(G, H)$ such that $AA^{-1} = I_G$ and $A^{-1}A = I_H$.
4. There exists A^* such that

$$\langle A\varphi, \psi \rangle = \langle \varphi, A^*\psi \rangle, \ \forall \varphi \in H, \psi \in G.$$

The operator A^* is called the **adjoint** of $A \in L(H, G)$.
5. An operator $A \in L(H, H) = L(H)$ is said to be **self-adjoint** if $A^* = A$. It is called **(strictly) positive** if

$$\langle A\varphi, \varphi \rangle \geqslant (>)0, \ \forall \varphi \in H.$$

6. An operator $U \in L(H, G)$ is said **unitary** if $U^*U = UU^* = I$.
7. One call **eigenvalues** $\lambda \in \mathbb{C}$ and **eigenfunctions** $\varphi \in H, \varphi \neq 0$, elements such that $A\varphi = \lambda\varphi$.
8. Define $A\xi$, the action of any operator $A \in L(G, H)$ on some Hilbert-space process $\xi : G \to L^2(\Omega, \mathscr{A}, P)$ by

$$\langle A\xi, \varphi \rangle = \langle \xi, A^*\varphi \rangle, \ \forall \varphi \in H.$$

Here are some standard results.

Lemma 1.2. 1. If $A \in L(H, G)$ and is bijective then A is invertible (i.e. A^{-1} is a linear bounded operator, $A^{-1} \in L(G, H)$).
2. If $A \in L(H, G)$ then $N(A) = R(A^*)^\perp$ and $\overline{R(A)} = N(A^*)^\perp$, where $\overline{(\cdot)}$ and $(\cdot)^\perp$ denote the closure and the orthogonal subspaces.
3. If $A \in L(H, G)$ then $A^* \in L(G, H)$.
4. If A is injective so is A^*A.
5. If $A \in L(H, G)$ then $A^*A \in L(H)$ is self-adjoint and positive.
6. A self-adjoint operator is injective if and only if its range is dense in H.
7. A self-adjoint operator is invertible if and only if $R(A) = H$.
8. A self-adjoint operator then

$$\|A\| = \sup_{\|\varphi\|=1} |\langle A\varphi, \varphi \rangle|.$$

9. If $U \in L(H,G)$ is unitary, then

$$\langle U\varphi, U\psi \rangle = \langle \varphi, \psi \rangle, \ \forall \varphi, \psi \in H.$$

Proof. (1) A proof may be found in [73].

(2) We have $\langle \varphi, A^*\psi \rangle = \langle A\varphi, \psi \rangle = 0$ for all $\varphi \in N(A), \psi \in G$. Hence, $N(A) = R(A^*)^\perp$. Interchanging the roles of A and A^* gives $N(A^*) = R(A)^\perp$. Thus, $N(A^*)^\perp = (R(A)^\perp)^\perp = \overline{R(A)}$.

(3) Straightforward.

(4) We have $\langle A^*A\varphi, \varphi \rangle = \langle A\varphi, A\varphi \rangle = \|A\varphi\|^2$. If $\varphi_0 \in N(A^*A)$ then $\varphi_0 \in N(A)$.

(5) We have $\langle A^*A\varphi, \psi \rangle = \langle A\varphi, A\psi \rangle = \langle \varphi, A^*A\psi \rangle$. Note also that $\langle A^*A\varphi, \varphi \rangle = \|A\varphi\|^2 \geqslant 0$.

(6) A injective if and only if $N(A) = \{0\}$ if and only if $N(A)^\perp = H$. We then use that $\overline{R(A)} = N(A^*)^\perp = N(A)^\perp$ by (2) and the fact that A is self-adjoint.

(7) By (6), A invertible is thus equivalent to $R(A) = H$.

(8) A proof may be found in [73].

(9) We have, since U is unitary,

$$\langle U\varphi, U\psi \rangle = \langle U^*U\varphi, \psi \rangle = \langle \varphi, \psi \rangle, \ \forall \varphi, \psi \in H.$$

Some new definitions and properties concerning mostly compact operators are presented here. Compact operators are very important in inverse problems for several reasons.

First, a compact operator is not invertible, i.e. has no bounded inverse (see Lemma 1.3). Thus if A is a compact operator the problem is naturally ill-posed in the sense of Definition 1.1. From a mathematical point of view, ill-posed problems are the more challenging.

Compact operators have simple spectra only composed of eigenvalues, see Theorem 1.1. This is a nice property of compact operators which gives rise to natural basis of functions to use, the singular value decomposition. By projection on this natural basis, we will obtain a sequence space model in Section 1.1.5. This model in the space of coefficients, is usually more easy to deal with from a statistical point of view.

Definition 1.4. 1. An operator A from H to G is called **compact** if each bounded set in H has an image by A which is relatively compact in G, i.e. with a compact closure.

2. Denote by $K(H,G)$ the space of **compact linear bounded operator**.

3. The **strong convergence**, denoted $\rightarrow_s$, is the convergence with respect to the norm in H or G.

4. The **weak convergence**, denoted $\rightarrow_w$, is the convergence with respect to $\langle \varphi, \cdot \rangle$ for all $\varphi \in H$ or G.

Lemma 1.3. *1. Let $A \in K(H,G)$, then there exist $A_n \in K(H,G)$, such that dim $R(A_n) < \infty$ and $\|A_n - A\| \to 0$, as $n \to \infty$.*
2. $A \in K(H,G)$ is equivalent to $A^* \in K(G,H)$
3. $A \in K(H,G)$ is equivalent to $\forall \varphi_k \in H : \varphi_k \to_w \varphi$ implies $A\varphi_k \to_s A\varphi$.
4. If $A \in K(H,G)$ and $dim(H) = \infty$ then A^{-1} is not bounded.
5. $A \in K(H)$ is equivalent to the fact that for any orthonormal sequence $\{\varphi_k\}$, one has $\lim_{k \to \infty} \langle A\varphi_k, \varphi_k \rangle = 0$.

Proof. A proof may be found in [73]. ∎

Theorem 1.1. *Let $A \in K(H)$ be self-adjoint. Then there exists a complete orthonormal system $E = \{\varphi_j : j \in I\}$ of H consisting of eigenfunctions of A. Here I is some index set and $A\varphi_j = \lambda_j \varphi_j$, for $j \in I$. The set $J = \{j \in I : \lambda_j \neq 0\}$ is countable and*

$$A\varphi = \sum_{j \in I} \lambda_j \langle \varphi, \varphi_j \rangle \varphi_j, \tag{1.3}$$

for all $\varphi \in H$. Moreover, for any $\delta > 0$ the set $J_\delta = \{j \in I : |\lambda_j| \geqslant \delta\}$ is finite.

Proof. This proof may be found in [72]. First, we prove the existence of an eigenvalue for a self-adjoint compact operator (if $H \neq \{0\}$). Due to Lemma 1.2, there exists a sequence $\{\varphi_k\}$ with $\|\varphi_k\| = 1$ such that, for $\lambda = \pm\|A\|$, $\langle A\varphi_k, \varphi_k \rangle \to \lambda$ as $k \to \infty$. Remark that

$$0 \leqslant \|A\varphi_k - \lambda\varphi_k\|^2 = \|A\varphi_k\|^2 - 2\lambda \langle A\varphi_k, \varphi_k \rangle + \lambda^2 \|\varphi_k\|^2$$

$$\leqslant \|A\|^2 - 2\lambda \langle A\varphi_k, \varphi_k \rangle + \lambda^2 \to 0, \text{ as } k \to \infty.$$

Thus, $A\varphi_k \to \lambda\varphi_k$ as $k \to \infty$. Since A is compact, there exists a subsequence such that $A\varphi_{k(n)} \to \psi$ as $n \to \infty$. It follows that $\lambda\varphi_{k(n)} \to \psi$ as $n \to \infty$. Denote $\varphi = \psi/\lambda$, therefore $\varphi_{k(n)} \to \psi$ as $n \to \infty$ and $A\varphi = \lambda\varphi$, since A is bounded.

We then prove that the system is orthogonal. Suppose $\lambda_j \neq \lambda_k$. We have

$$\langle A\varphi_j, \varphi_k \rangle = \lambda_j \langle \varphi_j, \varphi_k \rangle.$$

Moreover, since A is self-adjoint, we have

$$\langle A\varphi_j, \varphi_k \rangle = \langle \varphi_j, A\varphi_k \rangle = \lambda_k \langle \varphi_j, \varphi_k \rangle.$$

Thus φ_j and φ_k are orthogonal.

We now study the case where φ_j and φ_k are eigenfunctions with the same eigenvalue λ, suppose $\langle \varphi_j, \varphi_k \rangle = c \neq 0$. Thus, $\varphi_j - \varphi_k/c$ is still an eigenfunction related to λ and orthogonal to φ_j. One may easily orthonormalize the system.

The last part consists in proving the completness. By Zorn's Lemma, choose E the maximal set of eigenfunctions of A. Let S be the closed linear span of E. Obviously, $A(S) \subset S$. Moreover, $A(S^\perp) \subset S^\perp$, since $\langle As, \varphi \rangle = \langle s, A\varphi \rangle = 0$ for all $s \in S^\perp$ and all $\varphi \in S$. Remark that $A_{|S^\perp}$ is compact and self-adjoint. Hence, if $S^\perp \neq \{0\}$ there exists an eigenfunction $\psi \in S^\perp$ (by the first part of this proof). Since this contradicts the maximality of E, we conclude that $S^\perp = \{0\}$. Therefore the orthonormal system is complete. To show (1.3) we apply A to the representation

$$\varphi = \sum_{j \in I} \langle \varphi, \varphi_j \rangle \varphi_j. \tag{1.4}$$

Remark that only countable number of terms in (1.4) can be non-zero. Indeed, by Bessel's inequality we have

$$\sum_{\varphi_j \in E} |\langle \varphi, \varphi_j \rangle|^2 = \sup \left\{ \sum_{\varphi_j \in F} |\langle \varphi, \varphi_j \rangle|^2 : F \subset E, \mathrm{card}(F) < \infty \right\} \leqslant \|\varphi\|^2 < \infty.$$

Therefore, for any $k \in \mathbb{N}$, the set $S_k = \{\varphi_j \in E : |\langle \varphi, \varphi_j \rangle| \in [\|\varphi\|/(k+1), \|\varphi\|/k]\}$ is finite, and the union for all $k \in \mathbb{N}$ is then countable.

Assume that J_δ is infinite for some $\delta > 0$. Since A is compact, there exists a subsequence $\{\varphi_{k(n)}\}$ of $\{\varphi_k\}$ such that $\{A\varphi_{k(n)}\} = \{\lambda_{k(n)} \varphi_{k(n)}\}$ is a Cauchy sequence. This is in contradiction since $\|\lambda_k \varphi_k - \lambda_j \varphi_j\|^2 = \lambda_k^2 + \lambda_j^2 \geqslant 2\delta^2$ for $j \neq k$ due to the orthonormality of $\{\varphi_k\}$.

Remark 1.4. A linear bounded self-adjoint compact operator between two Hilbert spaces may thus be seen as an infinite matrix. In applications, a large matrix could be modelized by a compact operator. However, due to Theorem 1.1, the eigenvalues λ_j are going to 0. This is fundamental and characterizes the notion of ill-posed problems (see Definition 1.7). One observes a function through an operator A which, in some sense, concentrates to 0. Thus, the inversion of such an operator has to be made carefully, otherwise, the reconstruction will explose.

In general inverse problems, we neither assume that A is injective nor that $g \in R(A)$. Thus, we usually need some standard definitions of a generalized notion of inverse for the equation $Af = g$ (see [64]).

Definition 1.5. Let $A \in L(H, G)$.

1. We call f a **least-squares solution** of the problem (1.1) if

$$\|Af - g\| = \inf\{\|A\varphi - g\| : \varphi \in H\}.$$

2. We call f a **best approximate solution** of the problem (1.1) if it is a least-squares solution and if

$$\|f\| = \inf\{\|\varphi\| : \varphi \text{ is a least-squares solution}\}.$$

3. The **Moore-Penrose (generalized) inverse** $A^\dagger : D(A^\dagger) \to H$ of A defined on $D(A^\dagger) = R(A) \oplus R(A)^\perp$ maps $g \in D(A^\dagger)$ to the best approximate solution of (1.1). The existence of a best approximate solution is guaranted by $g \in R(A) \oplus R(A)^\perp$.

We have

Lemma 1.4. *Let $Q : G \to \overline{R(A)}$ be the orthogonal projection onto $\overline{R(A)}$. Then the three statements are equivalent:*

1. $f \in H$ *is a least-squares solution of (1.1).*
2. $Af = Qg$.
3. *The* **normal equation** $A^*Af = A^*g$ *holds.*

We have in addition the following properties for $g \in R(A) \oplus R(A)^{\perp}$.
4. *Any least-squares solution belongs to $A^{\dagger}g + N(A)$.*
5. *We also have that a best approximate solution exists, is unique and equals to $A^{\dagger}g$.*

Proof. Since Q is an orthogonal projection on $\overline{R(A)}$, remark that $\langle Af - Qg, (I - Q)g \rangle = 0$. We then have

$$\|Af - g\|^2 = \|Af - Qg\|^2 + \|(I - Q)g\|^2.$$

This shows that (2) implies (1). Vice versa, if f is a least-squares solution the last equation shows that f is a minimum of $\|Af - Qg\|$. Again by property of the projection we obtain (2).

Moreover, f is a least-squares solution if and only if Af is the closest element in $R(A)$ to g, which is equivalent to $Af - g \in R(A)^{\perp} = N(A^*)$, i.e. $A^*(Af - g) = 0$.

(4) Suppose that $g \in R(A) \oplus R(A)^{\perp}$. Then $Qg \in R(A)$ and (2) is true and there exists at least one least-squares solution f_0. Moreover, due to (2), any element of $f_0 + N(A)$ is also a least-squares solution.

(5) Remark that for any $u \in N(A)$:

$$\|f_0 + u\|^2 = \|(I - P)(f_0 + u)\|^2 + \|P(f_0 + u)\|^2 = \|(I - P)f_0\|^2 + \|Pf_0 + u\|^2,$$

where f_0 is a least-squares solution, P is the orthogonal projection on $N(A)$. This yields the uniqueness of the best approximate solution, which is equal to $(I - P)f_0$.

Obviously, if $A^{-1} \in L(G, H)$ exists then $A^{-1} = A^{\dagger}$.

Under assumptions of Lemma 1.4, the best approximate solution is in fact the least-squares solution with a null term in the null-space of A. Indeed, any $f \in N(A)$ is such that $Af = 0$, and cannot be observed through A. Thus, there is no real meaning in trying to reconstruct it.

The normal equation is a different way to express an inverse problem. Indeed one may multiply the first problem by A^* and then get the equivalent normal equation.

Remark 1.5. In a statistical inverse problem, we observe a (random) noisy version of Af. Thus, if A is injective then the unique best approximate solution is f (by Lemma 1.4).

1.1.5 Singular Value Decomposition and Sequence Space Model

Let $A \in L(H, G)$ be an injective and compact operator. We have, by applying Theorem 1.1 to A^*A, which is self-adjoint and strictly positive,

$$A^*Af = \sum_{k=1}^{\infty} \rho_k \langle f, \varphi_k \rangle \varphi_k,$$

where $\rho_k > 0$. Define the normalized image $\{\psi_k\} \in G$ of $\{\varphi_k\} \in H$ by

$$\psi_k = b_k^{-1} A \varphi_k,$$

where $b_k = \sqrt{\rho_k} > 0$. Remark that $\{\psi_k\}$ are orthogonal,

$$\langle \psi_k, \psi_\ell \rangle = b_k^{-1} b_\ell^{-1} \langle A\varphi_k, A\varphi_\ell \rangle = b_k^{-1} b_\ell^{-1} \langle A^*A\varphi_k, \varphi_\ell \rangle = b_k b_\ell^{-1} \langle \varphi_k, \varphi_\ell \rangle = \delta_{k\ell},$$

where $\delta_{k\ell}$ denotes the Kronecker symbol (0 if $k \neq \ell$, 1 if $k = \ell$). Note that this implies $\|\psi_k\|^2 = 1$. Thus, $\{\psi_k\}$ is an orthonormal system. Moreover

$$A^* \psi_k = b_k^{-1} A^* A \varphi_k = b_k^{-1} b_k^2 \varphi_k = b_k \varphi_k.$$

Thus, we have

$$A\varphi_k = b_k \psi_k, \quad A^* \psi_k = b_k \varphi_k.$$

The $b_k > 0$ are called **singular values** of the operator A. Note also that, since A^*A is compact and self-adjoint then $b_k \to 0$ as $k \to \infty$ by Theorem 1.1.

Definition 1.6. We say that A admits a **singular value decomposition (SVD)** if, $\forall f \in H$,

$$A^*Af = \sum_{k=1}^{\infty} b_k^2 \theta_k \, \varphi_k,$$

where θ_k are the coefficients of f in the orthonormal basis $\{\varphi_k\} \in H$, $\{b_k\}$ are the singular values.

The SVD is the natural basis for A since it diagonalizes A^*A.

Now consider the projection of Y on $\{\psi_k\}$

$$\langle Y, \psi_k \rangle = \langle Af, \psi_k \rangle + \varepsilon \langle \xi, \psi_k \rangle = \langle Af, b_k^{-1} A\varphi_k \rangle + \varepsilon \xi_k$$

$$= b_k^{-1} \langle A^*Af, \varphi_k \rangle + \varepsilon \xi_k = b_k \theta_k + \xi_k,$$

where $\xi_k = \langle \xi, \psi_k \rangle$.

Since ξ is a white noise $\{\xi_k\}$ is a sequence of i.i.d. standard Gaussian random variables $\mathcal{N}(0,1)$ by Lemma 1.1.

Thus, under these assumptions, one has the equivalent discrete sequence observation model derived from (1.2):

$$y_k = b_k \theta_k + \varepsilon \xi_k, \quad k = 1, 2, \dots, \tag{1.5}$$

where y_k stands for $\langle Y, \psi_k \rangle$. This model is called the **sequence space model**. The aim here is to estimate the sequence $\theta = \{\theta_k\}$ from the observations $y = \{y_k\}$.

One can see the influence of the ill-posedness of the inverse problem when A is compact. Indeed, since b_k are the singular values of a compact operator, then $b_k \to 0$

as $k \to \infty$. Thus, when k increases the 'signal' $b_k \theta_k$ is weaker and it is clearly more difficult to estimate θ_k.

Another comment concerns the fact that the aim is to estimate $\{\theta_k\}$ and not $\{b_k \theta_k\}$. Thus, one really has to consider the inverses of the b_k, i.e., to invert the operator A.

For this reason, the following equivalent model to (1.5) is more natural

$$X_k = \theta_k + \varepsilon \sigma_k \xi_k, \quad k = 1, 2, \ldots, \tag{1.6}$$

where $X_k = y_k / b_k$, and $\sigma_k = b_k^{-1} > 0$. Note that $\sigma_k \to \infty$. In this model the aim is to estimate $\{\theta_k\}$ from $\{X_k\}$. When k is large the noise in X_k may then be very large, making the estimation difficult.

The sequence space model (1.5) or (1.6) for statistical inverse problems was studied in many papers, see [39, 95, 78, 32], among others.

Remark 1.6. For ill-posed inverse problems we have $b_k \to 0$ and $\sigma_k \to \infty$, as $k \to \infty$. We can see that ill-posed problems are more difficult than the direct problem. Indeed, when k is large, the noise $\varepsilon \sigma_k \xi_k$ will dominate. Thus, the estimation of $\{\theta_k\}$ from $\{X_k\}$ is more involved.

One can characterize linear inverse problems by the difficulty of the operator, i.e. with our notations, by the behaviour of the σ_k. If $\sigma_k \to \infty$, as $k \to \infty$, the problem is ill-posed.

Definition 1.7. An inverse problem is called **mildly ill-posed** if the sequence σ_k has a polynomial behaviour when k is large

$$\sigma_k \asymp k^\beta, \ k \to \infty,$$

and **severely ill-posed** if σ_k tends to infinity at an exponential rate

$$\sigma_k \asymp \exp(\beta k), \ k \to \infty,$$

where $\beta > 0$ is called the **degree of ill-posedness** of the inverse problem.

A special case of inverse problems is the **direct problem** where

$$\sigma_k \asymp 1, \ k \to \infty,$$

which corresponds to $\beta = 0$.

Here and later, $a_n \asymp b_n$ means that there exist $0 < c_1 \leqslant c_2 < \infty$ such that, $c_1 \leqslant a_n / b_n \leqslant c_2$, as $n \to \infty$.

Remark 1.7. One may also consider inverse problems which are more difficult than severely ill-posed, in the case where $\sigma_k \asymp \exp(\beta k^r)$, where $\beta > 0$ and $r \geqslant 1$.

Remark 1.8. There exist more general definitions of the degree of ill-posedness related to the noise structure, smoothness assumptions on f, smoothing properties of A (see [131, 103]). However, for the sake of simplicity, we prefer to deal with the simple notion defined above.

Remark 1.9. An important special case is the case where $A = I$. This corresponds to the **direct problem** where f is directly observed (with noise) with no inverse problem, i.e. without the need of inverting some operator A. In this case $\sigma_k \equiv 1$ and the model in (1.6) corresponds to the classical sequence space model in statistics. The model is then related to the Gaussian white noise model and is very close to nonparametric regression with $\varepsilon = n^{-1/2}$ (see Section 1.1.6.1).

1.1.6 Examples

Here are some examples of ill-posed problems where the SVD may be applied. In each case, the SVD can be explicitly computed.

Moreover, from a practical point of view, methods based on SVD are usually rather expensive in term of computations. For these reasons, many populars methods nowadays do not use explicitly the SVD.

On the other hand, even for these methods, the spectral domain is often used in order to deal with the theoretical accuracy of the methods.

1.1.6.1 Standard Gaussian White Noise

One of most classical model in nonparametric statistics is the Gaussian white noise

$$dY(t) = f(t)dt + \varepsilon dW(t), \; t \in [0,1], \tag{1.7}$$

where one observes $\{Y(t), t \in [0,1]\}$, f is an unknown function in $L^2[0,1]$, W is a Wiener process, $\varepsilon > 0$ is the noise level. One may check easily that dW corresponds to a white noise. Indeed, we obtain directly from the definition of integral against a Wiener process that for all $\varphi \in L^2[0,1]$,

$$\int_0^1 \varphi(t)dW(t) \sim \mathcal{N}\left(0, \int_0^1 |\varphi(t)|^2 dt\right).$$

We also obtain the property for the scalar product by the definition of the Wiener process

$$\mathrm{Cov}\left(\int_0^1 \varphi_1(t)dW(t), \int_0^1 \varphi_2(t)dW(t)\right) = \int_0^1 \varphi_1(t)\varphi_2(t)dt = \langle \varphi_1, \varphi_2 \rangle,$$

for all $\varphi_1, \varphi_2 \in L^2[0,1]$.

This model is a very specific inverse problem since, in this case, the operator is $A = I$ and $H = G = L^2[0,1]$. However, most of the results on inverse problems will apply in this framework. This model is often called a **direct problem**, since from our definition we have at our disposal direct observations and not indirect ones.

In this case, the sequence space model may be obtained by projecting on any orthonormal basis $\{\psi_k\} \in L^2[0,1]$. Doing so, one obtains

$$\int_0^1 \psi_k(t)dY(t) = \int_0^1 \psi_k(t)f(t)dt + \varepsilon \int_0^1 \psi_k(t)dW(t),$$

which is equivalent to

$$y_k = \theta_k + \varepsilon \xi_k, \ k = 1, \ldots,$$

where the θ_k are the coefficients of f in $\{\psi_k\}$ and

$$\xi_k = \int_0^1 \psi_k(t)dW(t) \sim \mathcal{N}(0,1)$$

with $\{\xi_k\}$ i.i.d. We then obtain a sequence space model where $b_k \equiv 1$.

It is well-known that the Gaussian white noise model defined in (1.7) is an idealized version of the more standard nonparametric regression

$$Y_i = f(X_i) + \xi_i, \ i = 1, \ldots, n, \tag{1.8}$$

where $(X_1, Y_1), .., (X_n, Y_n)$ are observed (we may assume $X_i \in [0,1]$), f is an unknown function in $L^2[0,1]$, and $\{\xi_i\}$ are i.i.d. zero-mean Gaussian random variables with variance σ^2.

The Gaussian white noise model may be understood as a large sample limit of nonparametric regression in (1.8). Indeed, by projecting (1.7) on the intervals $I_i = [(i-1)/n, i/n]$, $i = 1, \ldots, n$, one obtains

$$n \int_{I_i} dY(t) = n \int_{(i-1)/n}^{i/n} f(t)dt + n\varepsilon \int_{(i-1)/n}^{i/n} dW(t).$$

Thus, if f is smooth enough, and $\varepsilon^2 = \sigma^2/n$, one has an informal writting

$$Y_i \asymp f(i/n) + \xi_i, \ i = 1, \ldots, n,$$

where $\{\xi_i\}$ are i.i.d. zero-mean Gaussian random variables with variance σ^2.

This equivalence is proved in different frameworks and models (nonparametric regression, density, non-Gaussian noise ...) in [13, 107, 63, 113, 129].

Thus, under proper calibration, i.e. $\varepsilon^2 \asymp \sigma^2/n$, the asymptotics of model (1.8) as $n \to \infty$ and (1.7) as $\varepsilon \to 0$ are equivalent with the asymptotics of the latter being easier to derive.

In the inverse problem context, model (1.2) may be seen as an idealized version of the discrete sample model

$$Y_i = Af(X_i) + \xi_i, \ i = 1, \ldots, n, \tag{1.9}$$

where $(X_1, Y_1), .., (X_n, Y_n)$ are observed (we may assume $X_i \in [0,1]$), f is an unknown function in $L^2[0,1]$, A is an operator from $L^2[0,1]$ into $L^2[0,1]$, and ξ_i are i.i.d. zero-mean Gaussian random variables with variance σ^2.

1.1.6.2 Derivation

Another related example, which does not exactly correspond to our framework, but is very important, is the estimation of a derivative. Suppose that we observe

$$Y = f + \varepsilon \xi, \tag{1.10}$$

where $H = L^2[0,1]$, f is a 1−periodic C^β function in $L^2[0,1]$, i.e. β continuously differentiable, $\beta \in \mathbb{N}$ and ξ is a white noise. A standard problem in statistics is the estimation of the derivative $D^\beta f = f^{(\beta)}$ of f, or the function f itself when $\beta = 0$ (which corresponds to the previous section). This problem is studied for example in [47].

One may use here the Fourier basis $\varphi_k(x) = e^{2\pi i k x}$, $k \in \mathbb{Z}$. Denote by θ_k the Fourier coefficients of f,

$$\theta_k = \int_0^1 f(x) e^{2\pi i k x} dx$$

and note that

$$D^\beta (e^{2\pi i k \cdot})(x) = (2\pi i k)^\beta e^{2\pi i k x}.$$

It is well-known that we then have

$$f^{(\beta)} = \sum_{k=-\infty}^{\infty} (2\pi i k)^\beta \theta_k \varphi_k.$$

We have the following equivalent model in the Fourier domain

$$y_k = \theta_k + \varepsilon \xi_k, \ k \in \mathbb{Z} \setminus \{0\},$$

and we want to estimate $v_k = \theta_k (2\pi i k)^\beta$. This is equivalent to, observing

$$y_k = (2\pi i k)^{-\beta} v_k + \varepsilon \xi_k, \ k \in \mathbb{Z} \setminus \{0\},$$

and estimating θ_k.

Thus, derivation is a mildly ill-posed inverse problem of degree β.

1.1.6.3 Circular Deconvolution

The framework of (circular) deconvolution is perhaps one of the most well-known inverse problem. It is used in many applications as econometrics, physics, astronomy, medical image processing. For example, it corresponds to the problem of a blurred signal that one wants to recover from indirect data.

Example 1.1. One famous example of an inverse problem of deconvolution is the blurred images of the Hubble space telescope. In the early 1990, the Hubble satellite was launched into low-earth orbit outside of the disturbing atmosphere in order to provide images with a spatial resolution never achieved before. Unfortunately,

quickly after launch, a manufacturing error in the main mirror was detected, causing severe spherical aberrations in the images. Therefore, before the space shuttle
Endeavour visited the telescope in 1993 to fix the error, astronomers employed inverse problem techniques to improve the blurred images (see [1]).

Consider the following convolution operator:

$$Af(t) = r * f(t) = \int_0^1 r(t-x)f(x)dx, \quad t \in [0,1],$$

where r is a known 1-periodic symmetric around 0 real-valued convolution kernel
in $L^2[0,1]$. In this model, A is a linear bounded operator from $L^2[0,1]$ to $L^2[0,1]$.

This operator is a Hilbert-Schmidt integral operator and it is an **Hilbert-Schmidt
operator**, i.e., it is such that for some (and then any) orthonormal basis $\{e_k\}$ we have
$\sum \|Ae_k\|^2 < \infty$. It is then a compact operator.

Remark that, if $\{\varphi_k\}$ and $\{\psi_k\}$ are the SVD bases defined in Section 1.1.5 then

$$\sum \|A\varphi_k\|^2 = \sum \langle A^*A\varphi_k, \varphi_k \rangle = \sum b_k^2 < \infty.$$

This shows that the singular values are decreasing rather fastly in this situation.

By simple computations one may see that the adjoint A^* is also a Hilbert-Schmidt
integral operator, with kernel $\overline{r(x-t)}$, where $\overline{(\cdot)}$ denotes the complex conjugate.
Since r is real-valued and symmetric around 0, the operator A is also self-adjoint.

Define then the following model

$$Y(t) = r * f(t) + \varepsilon \, \xi(t), \quad t \in [0,1], \tag{1.11}$$

where Y is observed, f is an unknown periodic function in $L^2[0,1]$ and $\xi(t)$ is a
white noise.

This model is quite popular and has been studied in a large number of statistical
papers, see [50, 39, 45, 32, 29].

Define here $\{\varphi_k(t)\}$ the real trigonometric basis on $[0,1]$:

$$\varphi_1(t) \equiv 1, \quad \varphi_{2k}(t) = \sqrt{2}\cos(2\pi kt), \quad \varphi_{2k+1}(t) = \sqrt{2}\sin(2\pi kt), \quad k = 1,2,\ldots.$$

A function in $L^2[0,1]$ may be decomposed on $\{\varphi_k(t)\}$.

Remark now that by a simple change of variables

$$\int_0^1 r(t-x)e^{2\pi ikx}dx = e^{2\pi ikt}\int_{-t}^{1-t} r(-y)e^{2\pi iky}dy = e^{2\pi ikt}\int_0^1 r(x)e^{2\pi ikx}dx,$$

by periodicity.

The SVD basis is then clearly here the Fourier basis, i.e. $e^{2\pi ik\cdot}$.

We make the projection of (1.11) on $\{\varphi_k(t)\}$, in the Fourier domain, and obtain

$$y_k = b_k\theta_k + \varepsilon\xi_k,$$

where $b_k = \sqrt{2} \int_0^1 r(x) \cos(2\pi kx)dx$ for even k, $b_k = \sqrt{2} \int_0^1 r(x) \sin(2\pi kx)dx$ for odd k, θ_k are the Fourier coefficients of f, and ξ_k are i.i.d. standard Gaussian random variables.

1.1.6.4 Heat Equation

Consider the following heat equation which describes the heat at time t and position x based on some initial conditions:

$$\frac{\partial}{\partial t}u(x,t) = \frac{\partial^2}{\partial x^2}u(x,t), \ u(x,0) = f(x), \ u(0,t) = u(1,t) = 0,$$

where $u(x,t)$ is defined for $x \in [0,1], t \in [0,T]$, and the initial condition f is a 1-periodic function. The problem is the following: given the temperature $g(x) = u(x,T)$ at time T find the initial temperature $f \in L^2[0,1]$ at time $t = 0$.

Due to the boundary conditions, one uses here the sine basis $(\{\sqrt{2}\sin(k\pi\cdot)\})$. Let $\theta_k(t) = \sqrt{2} \int_0^1 \sin(k\pi x)f(x)dx$ denote the Fourier coefficients of f with respect to the complete orthonormal system $\{\varphi_k\}$ of $L^2[0,1]$.

In this case, one obtains an ordinary differential equation in the Fourier domain, which provides the following expression for u:

$$u(x,t) = \sqrt{2} \sum_{k=1}^{\infty} \theta_k e^{-\pi^2 k^2 t} \sin(k\pi x).$$

The problem is then: given the final temperature $u(x,T) = Af(x)$ to find the initial temperature f. Thus, we may write this problem as an inverse problem with the operator

$$Af(x) = \int_0^1 \sum_{k=1}^{\infty} e^{-\pi^2 k^2 T} 2\sin(k\pi x) \sin(k\pi y)f(y)dy.$$

Thus, A is a linear bounded injective compact operator, whose SVD is given by the sine basis. The singular values b_k are equal to $e^{-\pi^2 k^2 T/2}$ and the problem is therefore severely ill-posed.

The model is then the following:

$$Y(x) = u(x,T) + \varepsilon \, \xi(x), \ x \in [0,1],$$

where ξ is a white noise in $L^2[0,1]$. We want here to recover $f \in L^2[0,1]$.

From a statistical point of view, the problem is the following: given a noisy version of the final temperature (at time T) find the unknown initial condition f (at time 0).

This framework has been studied in [61] and [24].

By projection on the sine basis, one obtains the following sequence space model:

$$y_k = b_k \theta_k + \varepsilon \xi_k,$$

where $b_k = e^{-\pi^2 k^2 T/2}$, θ_k are the Fourier (sine) coefficients of f, and ξ_k are i.i.d. standard Gaussian random variables.

Remark that here the problem is very difficult. Indeed, it is even worse than a severely ill-posed problem since the singular values are decreasing faster than exponentially.

From a practical point of view, one can see that an error of order 10^{-8} in the fifth Fourier coefficient of $u(x,T)$ may lead to an error of $1000C$ in the initial temperature $f(x) = u(x,0)$.

One has to be very careful when solving this kind of inverse problem.

1.1.6.5 Computerized Tomography

Computerized tomography is used in medical image processing and has been studied for a long time, see [104]. In medical X-ray tomography one tries to have an image of the internal structure of an object. This image is characterized by a function f. However, there is no direct observations of f. Suppose that one observes the attenuation of the X-rays. Denote by I_0 and I_1 the initial and final intensities, by x the position on a given line L and by $\Delta I(x)$ the attenuation for a small Δx. One then has

$$\Delta I(x) = -f(x)I(x)\Delta x,$$

which corresponds from a mathematical point of view to

$$\frac{I'(x)}{I(x)} = -f(x),$$

and then by integration

$$\log(I_1) - \log(I_0) = \log\left(\frac{I_1}{I_0}\right) = -\int_L f(x)dx.$$

Thus observing I_1/I_0 is equivalent to the observation of $\exp(-\int_L f(x)dx)$. By measuring attenuation of X-rays, one observes cross section of the body.

From a mathematical point of view this problem corresponds to the reconstruction of an unknown function f in $\mathbb{R}^2$ (or in general $\mathbb{R}^d$) based on observations of its Radon transform Rf, i.e., of integrals over hyperplanes.

Let $B = \{x \in \mathbb{R} : \|x\| \leqslant 1\}$ be the unit ball in $\mathbb{R}^2$. Consider the integrals of a function $f : B \to \mathbb{R}$ over all the lines that intersect B. We parametrize the lines by the length $u \in [0,1]$ of the perpendicular from the origin to the line and by the orientation $s \in [0,2\pi)$ of this perpendicular with respect to the x-axis.

Suppose that the function f belongs to $L^1(B) \cap L^2(B)$. Define the Radon transform Rf of the function f by

$$Rf(u,s) = \frac{\pi}{2(1-u^2)^{\frac{1}{2}}} \int_{-\sqrt{1-u^2}}^{\sqrt{1-u^2}} f(u\cos s - t\sin s, u\sin s + t\cos s)dt, \qquad (1.12)$$

where $(u,s) \in S = \{(u,s) : 0 \leqslant u \leqslant 1,\ 0 \leqslant s < 2\pi\}$. With this definition, the Radon transform $Rf(u,s)$ is π times the average of f over the line segment (parametrized by (u,s)) that intersects B. It is natural to consider Rf as an element of $L^2(S,d\mu)$ where μ is the measure defined by $d\mu(u,s) = 2\pi^{-1}(1-u^2)^{\frac{1}{2}}du\,ds$. This measure μ is here in order to renormalize over lines.

In this case, the Radon operator R is a linear, bounded and compact operator from $L^2(B)$ into $L^2(S,d\mu)$.

The SVD of the Radon transform was well-studied, e.g. by [37, 104]. To introduce it, define the set of double indices $\mathcal{L} = \{\ell = (j,k) : j \geqslant 0, k \geqslant 0\}$. An orthonormal complex-valued basis for $L^2(B)$ is given by

$$\tilde{\varphi}_\ell(r,t) = \pi^{-\frac{1}{2}}(j+k+1)^{\frac{1}{2}}Z_{j+k}^{|j-k|}(r)e^{i(j-k)t},\ \ell = (j,k) \in \mathcal{L},\ (r,t) \in B, \quad (1.13)$$

where Z_a^b denotes the Zernike polynomial of degree a and order b. The corresponding orthonormal functions in $L^2(S,d\mu)$ are

$$\tilde{\psi}_\ell(u,s) = \pi^{-\frac{1}{2}}U_{j+k}(u)e^{i(j-k)s},\ \ell = (j,k) \in \mathcal{L},\ (u,s) \in S, \quad (1.14)$$

where $U_m(\cos s) = \sin((m+1)s)/\sin s$ are the Chebyshev polynomials of the second kind. We have $R\tilde{\varphi}_\ell = b_\ell\tilde{\psi}_\ell$, with the singular values

$$b_\ell = \pi^{-1}(j+k+1)^{-\frac{1}{2}},\ \ell = (j,k) \in \mathcal{L}. \quad (1.15)$$

Since we work with real functions, we identify the complex bases (1.13) and (1.14) with the equivalent real orthonormal bases $\{\varphi_\ell\}$, $\{\psi_\ell\}$ in a standard way,

$$\varphi_\ell = \begin{cases} \sqrt{2}\,\mathrm{Re}(\tilde{\varphi}_\ell) & \text{if } j > k, \\ \tilde{\varphi}_\ell & \text{if } j = k, \\ \sqrt{2}\,\mathrm{Im}(\tilde{\varphi}_\ell) & \text{if } j < k. \end{cases} \quad (1.16)$$

The problem of tomography in statistics is studied, for example, in [80, 83, 39, 21]. The model is the following

$$Y(u,s) = Rf(u,s) + \varepsilon\xi(u,s),\ (u,s) \in S,$$

where ξ is a white noise in $G = L^2(S,d\mu)$.

The SVD basis is known for the Radon transform. However, this basis is very difficult to compute. By projection on $\{\psi_\ell\}$, one obtains the equivalent sequence space model,

$$y_\ell = b_\ell\theta_\ell + \varepsilon\,\xi_\ell,\ \ell = (j,k),\ j \geqslant 0,\ k \geqslant 0,$$

where $\theta_\ell = \langle f, \varphi_\ell \rangle$, and ξ_ℓ are i.i.d. standard Gaussian random variables.

Remark 1.10. In tomography, the problem is mildly ill-posed, since the singular values have a polynomial behaviour. The exact degree of ill-posedness is a bit different, since the problem is ill-posed, but is also a problem of estimation of a function in two dimensions, which is known to be more difficult. One often considers that $\beta = 1/2$, due to (1.15).

There exist several models of tomography (X-rays tomography, positron emission tomography, discrete tomography, tomography in quantum physics and so on). The models each have their own specificities but are however all linked to the Radon operator.

1.1.7 Spectral Theory

In this section, we generalize the statistical study of inverse problems to the case of not only compact but also linear bounded operators. This extension is needed since there exist a lot of natural inverse problems where the operator is not compact, see for example the deconvolution on $\mathbb{R}$ in Section 1.1.7.2. In this situation, one needs other tools than the SVD. Moreover, the spectral Theorem and functional calculus may also be used in the case of compact operators.

1.1.7.1 The Spectral Theorem

The Halmos version of the spectral Theorem is convenient for the study of inverse problems (see [68]).

Theorem 1.2. *Let $A \in L(H)$ be a self-adjoint operator defined on a separable Hilbert space H. There exist a locally compact space S, a positive Borel measure Σ on S, a unitary operator $U : H \to L^2(\Sigma)$, and a continuous function $\rho : S \to \mathbb{R}$ such that*

$$A = U^{-1}M_\rho U, \tag{1.17}$$

where M_ρ is the multiplication operator $M_\rho : L^2(\Sigma) \to L^2(\Sigma)$ defined $M_\rho \varphi = \rho \cdot \varphi$.

Proof. A proof may be found in [125].

Remark 1.11. This fundamental result means that any self-adjoint linear bounded operator is similar to a multiplication in some L^2-space.

Remark 1.12. In the special case where A is a compact operator, a well-known version of the spectral theorem (see Theorem 1.1) states that A has a complete orthogonal system of eigenvectors $\{\varphi_k\}$ with corresponding eigenvalues ρ_k. This is a special case of (1.17) where $S = \mathbb{N}$, Σ is the counting measure, $L^2(\Sigma) = \ell^2(\mathbb{N})$, and $\rho(k) = \rho_k$.

Remark 1.13. If A is not self-adjoint then we use A^*A, where A^* is the adjoint of A.

Using the spectral theorem one obtains the equivalent model to (1.2)

$$UY = U(Af + \varepsilon\xi) = UAf + \varepsilon U\xi = UU^{-1}M_\rho Uf + \varepsilon U\xi,$$

which gives in the spectral domain

$$Z = \rho \cdot \theta + \varepsilon \eta, \tag{1.18}$$

where $Z = UY$, $\theta = Uf$ and $\eta = U\xi$ is a white noise in $L^2(\Sigma)$ since U is a unitary operator. Indeed, we have the following lemma.

Lemma 1.5. *Let ξ be a white noise in G and $\eta = U\xi$ where U is a unitary operator. We have for all $\theta = Uf$ and $v = Uh$, where $f, h \in H$,*

$$\langle \eta, \theta \rangle \sim \mathcal{N}(0, \|\theta\|^2),$$

and

$$\mathbf{E}(\langle \eta, \theta \rangle \langle \eta, v \rangle) = \langle \theta, v \rangle.$$

Thus η is a white noise in $L^2(\Sigma)$.

Proof. For all $\theta = Uf$ with $f \in H$, we have

$$\langle \eta, \theta \rangle = \langle U\xi, Uf \rangle = \langle \xi, U^*Uf \rangle = \langle \xi, f \rangle \sim \mathcal{N}(0, \|f\|^2) = \mathcal{N}(0, \|\theta\|^2). \tag{1.19}$$

In the same way if $\theta = Uf$ and $v = Uh$, where $f, h \in H$, we have

$$\mathbf{E}(\langle \eta, \theta \rangle \langle \eta, v \rangle) = \mathbf{E}(\langle \xi, f \rangle \langle \xi, h \rangle) = \langle f, h \rangle = \langle \theta, v \rangle, \tag{1.20}$$

since U is unitary. Using (1.19) and (1.20), we obtain the lemma.

Remark 1.14. The model (1.18) really helps to understand the utility of spectral Theorem in inverse problems. Indeed, by use of the unitary transform U, one replaces the model (1.2), not always easy to handle with a general linear operator, by a multiplication by a function ρ. Moreover, since U is unitary the noise is still a white noise.

1.1.7.2 Deconvolution on $\mathbb{R}$

In this section, we present an example of application, see for example [49, 116] when the operator is not compact. The operator considered here is

$$Af(t) = r * f(t) = \int_{-\infty}^{\infty} r(t-u)f(u)du,$$

where $r * f$ denotes the convolution through a known filter $r \in L^1(\mathbb{R})$. The aim is to reconstruct the unknown function f.

Deconvolution is one of the most standard inverse problems. The problem of circular convolution, i.e. with a periodic kernel r on $[a, b]$, appears for example in [32] and in Section 1.1.6.3. The main difference is that for periodic convolution the operator is compact and the basis of eigenfunctions is the Fourier basis. It seems clear, from a heuristic point of view, that the results could be extended to the case of convolution on $\mathbb{R}$ by using the Fourier transform on $L^2(\mathbb{R})$ instead of the Fourier series. This heuristic extension can be made formal by resorting to the spectral Theorem (Theorem 1.2).

Suppose that r is a real-valued function symmetric around 0, then

$$\tilde{r}(\omega) = \int_{-\infty}^{\infty} e^{it\omega} r(t) dt = \int_{-\infty}^{\infty} \cos(t\omega) r(t) dt, \ \forall \omega \in \mathbb{R}.$$

Suppose also that $\tilde{r}(\omega) > 0$ for all $\omega \in \mathbb{R}$. It is straightforward to see that the operator A is self-adjoint and strictly positive, since r is real-valued and symmetric around 0 and $\tilde{r} > 0$.

Define the Fourier transform as a unitary operator from $L^2(\mathbb{R})$ into $L^2(\mathbb{R})$ by

$$(Ff)(\omega) = \frac{1}{\sqrt{2\pi}} \int_{-\infty}^{\infty} e^{it\omega} f(t) dt, \ \omega \in \mathbb{R}, \ f \in L^1(\mathbb{R}) \cap L^2(\mathbb{R}), \tag{1.21}$$

and its continuous extension on $L^2(\mathbb{R})$.

We have that

$$F(r * f)(\omega) = \tilde{r}(\omega).(Ff)(\omega);$$

hence $A = F^{-1} M_{\tilde{r}} F$.

The model is then the following

$$Y(t) = r * f(t) + \varepsilon \xi(t), \ \forall t \in \mathbb{R}$$

where $f \in L^2(\mathbb{R})$ is unknown, ξ is a white noise in $L^2(\mathbb{R})$.

By applying the Fourier transform we obtain

$$FY(\omega) = F(r * f)(\omega) + \varepsilon \, F\xi(\omega) = \tilde{r}(\omega).(Ff)(\omega) + \varepsilon \eta(\omega), \tag{1.22}$$

where, by Lemma 1.5, η is a white noise in $L^2(\mathbb{R})$.

1.1.7.3 Functional Calculus

In this section, the aim is to provide some important tools from operator theory linked to the spectral Theorem. Functional calculus is the main tool in order to modify the operator A, by applying functions to the operator. This result is crucial in the study of regularization methods. This section is based on [72].

Definition 1.8. Let $A \in L(H)$. The **resolvent** $\rho(A)$ of A is the set of all $\lambda \in \mathbb{C}$ for which $(\lambda I - A)$ is invertible. The **spectrum** of A is defined as $\sigma(A) = \mathbb{C} \setminus \rho(A)$.

Note that an eigenvalue is in the spectrum, but that not all points in the spectrum are eigenvalues. However, compact operators have spectra composed of eigenvalues.

It follows immediately that the spectrum is invariant by unitary transformations. Thus, $\sigma(A) = \sigma(M_\rho)$ with the notation of Theorem 1.2. One may also prove that $\sigma(A)$ is a closed and bounded set, hence is compact.

Lemma 1.6. *Let $A \in L(H)$ be self-adjoint.*

1. For any $f \in C(S)$, i.e. a continuous function on S, we have

$$\|M_f\| = \|f\|_{\infty, \text{supp}\Sigma},$$

where the norm is the restricted sup-norm on $\text{supp}\Sigma$,

$$\text{supp}\Sigma = S \setminus \bigcup_{V \text{open}, \Sigma(V)=0} V.$$

2. We have $\sigma(A) = \overline{\rho(\text{supp}\Sigma)}$.
3. We have $\sigma(A) \subset \mathbb{R}$ and $\forall \lambda \in \mathbb{C}$

$$\|(\lambda I - A)^{-1}\| \leqslant |\text{Im}\lambda|^{-1}.$$

4. Let

$$m_- = \inf_{\|\varphi\|=1} \langle A\varphi, \varphi \rangle, \text{ and } m_+ = \sup_{\|\varphi\|=1} \langle A\varphi, \varphi \rangle,$$

then $\sigma(A) \subset [m_-; m_+]$.
5. We have $\sigma(A^*A) \subset [0, \|A^*A\|]$.

Proof. See [73].

If $p(\rho)$ is a polynomial in ρ then it is natural to define

$$p(A) = \sum_{j=0}^{k} p_j A^j. \tag{1.23}$$

The next theorem generalizes the idea of applying continuous functions to the operator A by just applying continuous functions to the spectrum, i.e. in $C(\sigma(A))$.

Theorem 1.3. *With the notation of Theorem 1.2, define*

$$\Phi(A) = U^{-1} M_{\Phi \circ \rho} U, \tag{1.24}$$

for a continuous real-valued $\Phi \in C(\sigma(A))$, where $\Phi \circ \rho(\omega) = \Phi(\rho(\omega))$. Then $\Phi(A) \in L(H)$ is self-adjoint and satisfies (1.23) if Φ is polynomial. The mapping $\Phi \to \Phi(A)$ is called **functional calculus** *at A and is an isometric algebra homomorphism from $C(\sigma(A))$ to $L(H)$, i.e., for all $\Phi, \Psi \in C(\sigma(A))$ and $\alpha, \beta \in \mathbb{R}$ we have*

$$(\alpha\Phi + \beta\Psi)(A) = \alpha\Phi(A) + \beta\Psi(A), \tag{1.25}$$

$$(\Phi\Psi)(A) = \Phi(A)\Psi(A), \tag{1.26}$$

$$\|\Phi(A)\| = \|\Phi\|_\infty. \tag{1.27}$$

The functional calculus is uniquely determined by (1.23) and (1.25)-(1.27).

Proof. This proof may be found in [72]. By Lemma 1.6, $\Phi(A)$ is bounded. It is self-adjoint because Φ is real-valued and

$$(\Phi(A))^* = U^{-1}(M_{\Phi \circ \rho})^* U = \Phi(A).$$

For a polynomial p we have $p(A) = U^{-1} p(M_\rho) U$ with (1.23). Since by definition of the multiplication operator $p(M_\rho) = M_{p \circ \rho}$, this corresponds to definition (1.24). The proof of (1.25) is clear. Moreover, we have

$$\Phi(A)\Psi(A) = U^{-1} M_{\Phi \circ \rho} U U^{-1} M_{\Psi \circ \rho} U = U^{-1} M_{\Phi \Psi \circ \rho} U = (\Phi \Psi)(A).$$

Finally, by Lemma 1.6 and continuity of Φ we obtain

$$\|\Phi(A)\| = \|M_{\Phi \circ \rho}\| = \|\Phi \circ \rho\|_\infty = \|\Phi\|_{\infty, \sigma(A)}.$$

Let $\Phi_A : C(\sigma(A)) \rightarrow L(H)$ be any isometric algebra homomorphism satisfying $\Phi_A(p) = p(A)$ for all polynomials. By the Weierstrass approximation Theorem, for any $\Phi \in C(\sigma(A))$ there exists a sequence of polynomials p_k such that $\|\Phi - p_k\|_{\infty, \sigma(A)} \rightarrow 0$ as $k \rightarrow \infty$. Using the property for the norm we obtain the desired unicity

$$\Phi_A(\Phi) = \lim_{k \to \infty} \Phi_A(p_k) = \lim_{k \to \infty} p_k(A) = \Phi(A).$$

Remark 1.15. The functional calculus may be extended to the case of bounded functions on $\sigma(A)$. The isometry is then replaced by an upper bound in (1.27).

Remark 1.16. This theorem is a fundamental tool in analysis of inverse problems. It allows to apply functions to the operator and then to its spectrum. The aim is to study the behaviour of A^{-1} or of more stable inverses (e.g. the regularized inverse).

Example 1.2. Regularization methods. Suppose that $A \in L(H)$ is self-adjoint and positive. In ill-posed problems, if A is not invertible, then 0 is in the spectrum. Another way to understand this is by Theorem 1.3. Indeed, when 0 is in the spectrum then the function $\Phi(x) = 1/x$ is not even bounded on $\sigma(A) \subset [0, \|A\|]$, it exploses at point 0. Thus $\Phi(A) = A^{-1}$ is not a bounded operator by (1.27). There is a need to invert A in a more stable way. This is exactly the role of regularization methods. One way to invert A, is by a small modification of the function Φ. For example, one may use $\Phi_\gamma(x) = 1/(x + \gamma)$ where $\gamma > 0$, which is continuous and bounded on $\sigma(A)$. This is exactly the idea of the Tikhonov regularization method, see Section 1.2.2.

1.2 Nonparametric Estimation

The aim of nonparametric estimation is to estimate (reconstruct) a function f (density or regression funtion) based on some observations. The main difference with parametric statistics is that the function f is not in some parametric family of functions, for example, the family of Gaussian probability density functions $\{\mathcal{N}(\mu, 1), \ \mu \in \mathbb{R}\}$.

Instead of a general framework, the problem of nonparametric estimation will be described here in the setting of the sequence space model (1.5) which is related to the inverse problem with random noise (1.2).

1.2.1 Minimax Approach

Let $\hat{\theta} = (\hat{\theta}_1, \hat{\theta}_2, \dots)$ be an estimator of $\theta = (\theta_1, \theta_2, \dots)$ based on the data $X = \{X_k\}$. An estimator of θ may be any measurable function of the observation $X = \{X_k\}$.

Then f is estimated by $\hat{f} = \sum_k \hat{\theta}_k \varphi_k$, where $\{\varphi_k\}$ is a basis.

The first point is to define the accuracy of some given estimator $\hat{\theta}$. Since an estimator is by definition random, we will measure the squared difference between $\hat{\theta}$ and the true θ, and then take the mathematical expectation.

Define the **mean integrated squared risk (MISE)** of $\hat{f}$ by

$$\mathscr{R}(\hat{f}, f) = \mathbf{E}_f \|\hat{f} - f\|^2 = \mathbf{E}_\theta \sum_{k=1}^{\infty} (\hat{\theta}_k - \theta_k)^2 = \mathbf{E}_\theta \|\hat{\theta} - \theta\|^2,$$

where the second equality follows from Parseval's Theorem (and relies on the fact that $\{\varphi_k\}$ is a basis), where the notation $\|\cdot\|$ stands for ℓ^2-norm of θ-vectors in the sequence space. Here and in the sequel $\mathbf{E}_f$ and $\mathbf{E}_\theta$ denote the expectations w.r.t. Y or $X = (X_1, X_2, \dots)$ for models (1.2) and (1.5) respectively. Analyzing the risk $\mathscr{R}(\hat{f}, f)$ of the estimator $\hat{f}$ is equivalent to analyze the corresponding sequence space risk

$$\mathscr{R}(\hat{\theta}, \theta) = \mathbf{E}_\theta \|\hat{\theta} - \theta\|^2.$$

The aim would be to find the estimator with the minimum risk. However, the risk of an estimator depends, by definition, on the unknown f or θ.

To that end, we assume that f belongs to some class of function $\mathscr{F}$.

Definition 1.9. Define the **maximal risk** of the estimator $\hat{f}$ on $\mathscr{F}$ as

$$\sup_{f \in \mathscr{F}} \mathscr{R}(\hat{f}, f),$$

and the **minimax risk** as

$$r_\varepsilon(\mathscr{F}) = \inf_{\hat{f}} \sup_{f \in \mathscr{F}} \mathscr{R}(\hat{f}, f),$$

where the $\inf_{\hat{f}}$ is taken over all possible estimators of f.

It is usually not possible in nonparametric statistics to find estimators which attain the minimax risk. A more natural approach is to consider the asymptotic properties, i.e. when the noise level tends to 0 ($\varepsilon \to 0$).

Definition 1.10. Suppose that some estimator $\tilde{f}$ is such that there exist constants $0 < C_2 \leqslant C_1 < \infty$ with, as $\varepsilon \to 0$

$$\sup_{f \in \mathscr{F}} \mathscr{R}(\tilde{f}, f) \leqslant C_1 v_\varepsilon^2,$$

where the positive sequence v_ε is such that $v_\varepsilon \to 0$ and

$$\inf_{\hat{f}} \sup_{f \in \mathscr{F}} \mathscr{R}(\hat{f}, f) \geqslant C_2 v_\varepsilon^2.$$

In this case the estimator $\tilde{f}$ is said to be **optimal** or to **attain the optimal rate of convergence** v_ε^2.

In the special case where $C_1 = C_2$ the estimator $\tilde{f}$ is said to be **minimax** or to **attain the exact constant**.

An optimal estimator is then an estimator whose risk is of the order of the best possible estimator.

Minimax estimation in nonparametric statistics is nowadays a classical approach. It goes back to [11, 122] and also [74]. Since the 80's, it has been obtained in many different models (nonparametric regression, Gaussian white noise, density estimation, spectral density estimation), with a varied form of estimators (kernels, projections, splines, wavelets), and for most classes of functions (Besov, Hölder, Sobolev, ...).

These kind of results are often considered as a first step in order to prove that a given method has good theoretical properties. Indeed, one has a criterion, which garantees that on some class of functions a given method is optimal.

Minimax estimation for statistical inverse problems (1.2) (or for its sequence space analogue (1.5)) was discussed in a number of papers. Optimal rates of convergence in this problem are obtained in [80, 84, 39, 83, 95, 47, 78, 9] and in related frameworks in [52, 19].

Exact asymptotics of the minimax L^2-risks are known in the deconvolution problem with somewhat different setups [50], in inverse problems for partial differential equations [61] and in tomography, for minimax L^2-risks among linear estimators [80]. Exact asymptotics for pointwise risks on the classes of analytic functions in tomography are due to [20].

A recent discussion of the different rates may be found in the review [93].

Remark 1.17. We only use the standard L^2-risk along these notes. However, many results may be obtained with different risks and loss functions, for example, the L^p, L^∞ or the pointwise risk, see [43, 85, 20, 79].

1.2.2 Regularization Methods

1.2.2.1 Continuous Regularization Methods

The main part of ill-posed inverse problems is to find regularization methods which will help to get a fine reconstruction of f.

Recall that the normal equation, defined in Lemma 1.4, is

$$A^*Y = A^*Af + \varepsilon A^*\xi.$$

Formally, one has to estimate the solution $(A^*A)^{-1}A^*Y$. The problem in ill-posed situation is that the operator A^*A is not (boundedly) invertible.

The idea is to get some continuous inversion by use of regularization methods. This allows to obtain much more stable reconstruction.

Definition 1.11. We call a **regularization method** an estimator defined by

$$\hat{f}_\gamma = \Phi_\gamma(A^*A)A^*Y,$$

where $\Phi_\gamma \in C(\sigma(A^*A))$, i.e a continuous function on $\sigma(A^*A)$ (or even bounded) depending on some **regularization parameter** $\gamma > 0$.

We are going to give some examples of regularization methods or estimators which are commonly used. All these methods are defined in the spectral domain even if some of them may be computed without using the whole spectrum.

Spectral Cut-Off

This regularization method is very simple. The idea is to get rid of the high frequencies. In the spectral domain, by using the spectral cut-off, one just cut the frequencies over some threshold.

The definition of a **spectral cut-off** with parameter $\gamma > 0$ is the following

$$\Phi_\gamma(x) = \begin{cases} x^{-1}, & x \geqslant \gamma, \\ 0, & x < \gamma. \end{cases}$$

This notion may be well-defined by use of the functional calculus for bounded functions (instead of continuous ones).

The spectral cut-off is a very simple estimator. It is usually used as a benchmark since it attains the optimal rate of convergence. However, it is not a very precise estimator. Moreover, from a numerical point of view, it is usually time consuming since one has to compute the whole spectrum.

Tikhonov Regularization

The Tikhonov method is one of the first and the most well-known regularization method in inverse problems.

The direct inversion of the operator A^*A is not satisfying since it is not a (boundedly) invertible operator. The idea is to control the norm of the solution by using a penalty term.

Define now, the well-known **Tikhonov regularization method** (see [127]). In this method one wants to minimize the following functional $L_\gamma(\varphi)$:

$$\inf_{\varphi \in H} \left\{ \|A\varphi - Y\|^2 + \gamma \|\varphi\|^2 \right\}, \tag{1.28}$$

where $\gamma > 0$ is some tuning parameter.

The Tikhonov method is very natural. Indeed, the idea is to choose an estimator which, due to the first term will fit the data, and which will be "stable", due to the second term, which is called the energy. As we will see in Section 1.3.3 the choice of γ is very sensitive since it characterizes the balance between the fitting and the stability.

The functional L_γ is strictly convex for any $\gamma > 0$. Its minimum is attained when its differential in $h \in H$

$$(L_\gamma)'_\varphi h = 2\langle A\varphi - Y, Ah \rangle + 2\gamma \langle \varphi, h \rangle, \tag{1.29}$$

is zero, i.e.

$$\langle A^*(Y - A\varphi), h \rangle = \gamma \langle \varphi, h \rangle, \ \forall h \in H.$$

The minimum is then attained by

$$\hat{f}_\gamma = (A^*A + \gamma I)^{-1} A^* Y. \tag{1.30}$$

In the spectral domain this method is defined by

$$\Phi_\gamma(x) = \frac{1}{x + \gamma}.$$

Remark 1.18. There exist some troubles with this simple Tikhonov regularization. For these reasons, several modifications of the Tikhonov have been defined.

Variants of Tikhonov Regularization

There exist several modified versions of the Tikhonov regularization.

The first variant is the **Tikhonov method with starting point** φ_0. It consists in giving a different starting point than 0. We have $\varphi_0 \in H$ and we penalize by $\|\varphi - \varphi_0\|$ instead of $\|\varphi\|$. By use of (1.29) one then obtains

$$\hat{f}_\gamma = (A^*A + \gamma I)^{-1}(A^*Y + \gamma \varphi_0). \tag{1.31}$$

The second modified version, which already appears in [127], is the **Tikhonov method with a different prior**. It is is based on the idea that the function could be smoother. Thus, a penalty term of the form $\|Q^a \varphi\|$, where $Q^a, a > 0$, is some differential operator, would be more suited. A classical example is then $Q^a = (A^*A)^{-a}$. The estimator is then defined by

$$\hat{f}_\gamma = (A^*A + \gamma (Q^a)^* Q^a)^{-1} A^* Y.$$

With this method one is able to better estimate smoother functions. Indeed, the standard Tikhonov method, penalize only by $\|\varphi\|^2$. If the function is smoother, then it is natural to take this into account in the second term, by a smoothness constraint. This effect may be seen in the better qualification of the method (see Section 1.2.3.1).

A last variant is called **iterative Tikhonov method**. It consists in starting a first Tikhonov regularization with $\varphi_0 = 0$ and then obtain an estimator $\hat{f}_\gamma^1$. In the second iteration, one applies the Tikhonov method with a starting point $\hat{f}_\gamma^1$. We iterate this method several times. The estimate is then by (1.31)

$$\hat{f}_{m+1} = (A^*A + \gamma I)^{-1}(A^*Y + \gamma \hat{f}_m).$$

It may be shown by induction that

$$\hat{f}_m = (A^*A + \gamma I)^{-m}(A^*A)^{-1}((A^*A + \gamma I)^m - \gamma^m I)A^*Y.$$

For $m = 1$ this corresponds exactly to the standard Tikhonov regularization. With this method one is able to better estimate smoother functions, the qualification of the method is increased (see Section 1.2.3.1). From a numerical point of view, this method is not really much longer than the Tikhonov one, since the only operator to invert is $(A^*A + \gamma I)$.

In the spectral domain this method is defined by

$$\Phi_\gamma(x) = \frac{(x + \gamma)^m - \gamma^m}{x(x + \gamma)^m}.$$

Landweber Iteration

Another very standard method is based on the idea to minimize the functional $\|A\varphi - Y\|$ by the steepest descent method (i.e. Gradient descent algorithm). The idea then is to choose the direction h equals to minus the gradient (in fact here the approximate gradient). Thus, we obtain $h = -A^*(A\varphi - Y)$ by (1.29). This leads to the recursion formula $\varphi_0 = \hat{f}_0 = 0$ and

$$\hat{f}_m = \hat{f}_{m-1} - \mu A^*(A\hat{f}_{m-1} - Y),$$

for some $\mu > 0$. This method is called **Landweber iteration**, see for example in [86].

It may be shown by induction that

$$\hat{f}_m = \sum_{j=0}^{m-1}(I - \mu A^*A)^j \mu A^*Y.$$

Indeed, it is clearly true for $m = 0$ and if true for m then

$$\hat{f}_{m+1} = (I - \mu A^* A)\hat{f}_m + \mu A^* Y = \sum_{j=0}^{m}(I - \mu A^* A)^j \mu A^* Y.$$

The parameter μ has to be chosen such that $\mu \|A^* A\| \leqslant 1$ which has a strong influence on the speed of convergence. The regularization parameter is then linked to the number of iterations m. Formally, the number of iterations may be written as γ^{-1}.

In the spectral domain this method is defined, for $\mu = 1$ and $\|A^* A\| = 1$, by

$$\Phi_m(x) = \sum_{j=0}^{m-1}(1 - x)^j.$$

There exist another version of this formula which will be used later. We have

$$\Phi_m(x) = \frac{1 - (1 - x)^m}{x}, \quad (\Phi_m(0) = m).$$

Remark 1.19. From a numerical point of view, this method is faster than Tikhonov method, since one does not need here the inversion of an operator (as in (1.30)).

However, Landweber iteration has some drawbacks as we will see in Section 1.2.3.1. Indeed, the number of iterations may be very large. For this reason, new methods, based on Landweber have been defined, as the semi-iterative procedures and ν-methods.

Semi-iterative Procedures and ν-Methods

As we will see the Landweber iteration is not so efficient. One of the reason is, that this method uses only the previous iteration in order to compute the next one. A more general idea is to use all the previous iterations $\hat{f}_j$, $j = 1, \dots, m-1$, to define $\hat{f}_m$.

This is the starting point of the so-called **semi-iterative procedures**. Let $\hat{f}_j$, $j = 1, \dots, m-1$, and $\hat{f}_0 = 0$ then define

$$\hat{f}_m = \mu_{1,m}\hat{f}_{m-1} + \cdots + \mu_{m,m}\hat{f}_0 + \omega_m A^*(Y - A\hat{f}_{m-1}),$$

where $\sum_j \mu_{j,m} = 1$.

The semi-iterative methods are then defined by

$$\hat{f}_m = \Phi_m(A^* A)A^* Y,$$

where Φ_m is a polynomial of degree exactly $m-1$, which is called iteration polynomial.

Clearly, such a method is computationaly rather efficient but we use all the iterations and not only one.

A special case of such iterative method are the ν-**methods** which only use two iterations. These methods were introduced in [10] and in the statistical literature by [106]. It is defined as a semi-iterative procedure with a parameter $v > 0$,

$$\mu_1 = 1, \ \omega_1 = \frac{4v+2}{4v+1},$$

$$\mu_m = 1 + \frac{(m-1)(2m-3)(2m+2v-1)}{(m+2v-1)(2m+4v-1)(2m+2v-3)},$$

$$\omega_m = 4\frac{(2m+2v-1)(m+v-1)}{(m+2v-1)(2m+4v-1)},$$

and

$$\hat{f}_m = \mu_m \hat{f}_{m-1} + (1-\mu_m)\hat{f}_{m-2} + \omega_m A^*(Y - A\hat{f}_{m-1}).$$

Remark 1.20. We will see in Section 1.2.3.1 that v-methods, and many semi-iterative methods, are much faster than the Landweber method. We will explain, what is the idea behind these v-methods.

Risk of Regularization Methods

A regularization method defined by Φ_γ may be decomposed as

$$\hat{f}_\gamma = \Phi_\gamma(A^*A)A^*Af + \varepsilon\Phi_\gamma(A^*A)A^*\xi, \tag{1.32}$$

since (1.2). Its risk may then be written as

$$\mathbf{E}_f\|\hat{f}_\gamma - f\|^2 = \|\mathbf{E}_f(\hat{f}_\gamma) - f\|^2 + \mathbf{E}_f\|\hat{f}_\gamma - \mathbf{E}_f(\hat{f}_\gamma)\|^2,$$

since $\mathbf{E}\Phi_\gamma(A^*A)A^*\xi = 0$. The first term is called the **approximation error** and the second is called **propagated noise error**.

Remark that by Theorem 1.3 we have

$$\Phi_\gamma(A^*A)A^*Af = U^{-1}M_{\Phi_\gamma(\rho)\rho}Uf.$$

Thus, $\Phi_\gamma(A^*A)$ should be an approximate inverse of A^*A. The study of the function $\Phi_\gamma(x)x$ in the spectral domain is then of major importance.

Remark 1.21. As for the estimation method, a key-point in regularization methods is to choose the parameter γ in a proper way.

1.2.2.2 Estimation Procedures

Equivalence in the Sequence Space Model

In order to get a framework more standard in statistics, suppose now that the operator A is compact. Then, by using the SVD, one obtains the sequence space model (1.5). Usually statisticians prefer to work with the sequence space model (1.6).

In this context, many regularization methods may be expressed in a statistical framework, and usually correspond to some known estimation method in statistics. The notion of regularization is not really used in statistics. However, there exists a more standard definition which is related.

Let $\lambda = (\lambda_1, \lambda_2, \ldots)$ be a sequence of nonrandom weights. Every sequence λ defines a **linear estimator** $\hat{\theta}(\lambda) = (\hat{\theta}_1, \hat{\theta}_2, \ldots)$ where

$$\hat{\theta}_k = \lambda_k X_k \text{ and } \hat{f}(\lambda) = \sum_{k=1}^{\infty} \hat{\theta}_k \varphi_k.$$

Remark also by use of the SVD in Theorem 1.1 or Theorem 1.3 one obtains for a general regularization method

$$\hat{f}(\lambda) = \Phi_\gamma(A^*A)A^*Y = \sum_{k=1}^{\infty} \Phi_\gamma(b_k^2) b_k y_k \varphi_k = \sum_{k=1}^{\infty} \Phi_\gamma(b_k^2) b_k^2 X_k \varphi_k,$$

which exactly corresponds to the special case of linear estimator with

$$\lambda_k = \Phi_\gamma(b_k^2) b_k^2. \tag{1.33}$$

Truncated SVD

Examples of commonly used weights λ_k are the projection weights $\lambda_k = I(k \leqslant N)$ where $I(\cdot)$ denotes the indicator function. These weights correspond to the **projection estimator** (also called **truncated SVD**).

$$\hat{\theta}(N) = \begin{cases} X_k, & k \leqslant N, \\ 0, & k > N. \end{cases}$$

The value N is called the **bandwidth**.

The projection estimator is then defined by

$$\hat{f}_N = \sum_{k=1}^{N} X_k \varphi_k.$$

The truncated SVD is a very simple estimator. With this natural estimator, one estimates the first N coefficients θ_k by their empirical counter-part X_k and then estimate the remainder terms by 0 for $k > N$.

This is an estimator equivalent to the spectral cut-off, but expressed in a different way and in a different setting. From a numerical point of view, it is still usually time consuming since, one has to compute all the coefficients X_k.

One may easily check by using (1.33) that, for the case $\sigma_k = k^\beta$, the spectral cut-off is equivalent to the projection estimator with $N = [\gamma^{-1/2\beta}]$.

Kernel Estimator

One of the most well-known method in statistics is the **kernel estimator** (see [115, 129]). In our context, kernel estimator could be defined in the special case of the direct problem; i.e. $A = I$. A kernel estimator is defined by its kernel function $K \in L^2$ (usually also $K \in L^1$) and

$$\hat{f}_\gamma = K_\gamma * Y,$$

where $*$ denotes the convolution product, $K_\gamma(\cdot) = \gamma^{-1} K(\cdot/\gamma)$ and $\gamma > 0$ is known as the bandwidth.

The idea of kernel estimators is to estimate the function f by using a local (by the bandwidth) weighted mean of the data, i.e. a convolution.

Kernel estimators may also be defined in inverse problem framework in order to invert the operator, see for example the so-called deconvolution kernel in [52].

This method is also linked to the mollifier methods in inverse problems, see [94].

The Tikhonov Estimator

The **Tikhonov estimator** is defined by the same minimization in (1.28) as for the Tikhonov regularization. In a more statistical framework, one may define the Tikhonov estimator by its equivalent form in the SVD domain:

$$\lambda_k = \frac{1}{1 + \gamma \sigma_k^2},$$

which is easy to verify by use of (1.33).

In the special case where $A = I$ this estimator is defined and computed as a modified version of the Tikhonov regularization and is called spline (see [131]).

In the parametric context of the standard linear regression, this method is called ridge regression, see [70]. It is known to improve on the standard least-squares estimator when the singular values of the design matrix are close to 0.

The Landweber Method

The Landwber iteration is not really known under this name in statistics. However, there exists a well-known approach in the community of learning which is strongly related.

Boosting algorithms include a family of iterative procedures which improve the performance at each step. The L^2-boosting has been introduced in the context of regression and classification in [15].

The idea is to start from a weak learner, i.e. a rather rough estimator $\hat{f}_0$. The algorithm consists then in boosting this learner in a recursive iteration, which may be showed to correspond to Landweber iteration (see [9]).

The Pinsker Estimator

The Pinsker estimator has been defined in [109]. This special class of linear estimators is defined in the sequence space model by the following weights coefficients

$$\lambda_k = (1 - c_\varepsilon a_k)_+,$$

where c_ε is the solution of the equation

$$\varepsilon^2 \sum_{k=1}^{\infty} \sigma_k^2 a_k (1 - c_\varepsilon a_k)_+ = c_\varepsilon L,$$

with $x_+ = \max(0,x)$ and $a_k > 0$.

As we will see in Section 1.2.4, this class of estimators is defined in the context of estimation in ellipsoids where they attain the optimal rates of convergence, but also the minimax constants.

Risk of a Linear Estimator

Define now the L^2–risk of linear estimators :

$$\mathscr{R}(\hat{\theta}(\lambda), f) = R(\theta, \lambda) - \mathbf{E}_\theta \sum_{k=1}^{\infty} (\hat{\theta}_k(\lambda) - \theta_k)^2 = \sum_{k=1}^{\infty} (1 - \lambda_k)^2 \theta_k^2 + \varepsilon^2 \sum_{k=1}^{\infty} \sigma_k^2 \lambda_k^2.$$
$$(1.34)$$

The first term in the RHS is called **bias term** and the second term is called the **stochastic term** or **variance term**. The bias term is linked to the approximation error and measure if the chosen estimation procedure is a good approximation of the unknown f. On the other hand, the stochastic term measure the influence of the random noise and of the inverse problem in the accuracy of the method.

In these lectures, we are going to study in details the projection estimators. This method is the most simple one and can be studied in a very easy way. The risk of a projection estimator with bandwidth N is

$$R(\theta, N) = \sum_{k=N+1}^{\infty} \theta_k^2 + \varepsilon^2 \sum_{k=1}^{N} \sigma_k^2.$$

In this case the decomposition is very simple. Indeed, we estimate the first N coefficients by their empirical version X_k and the other coefficients by 0. Thus, the bias term measure the influence of the remainder coefficients θ_k, $k > N$, and the stochastic term is due to the random noise in the N first coefficients. We can see now that one simple question is how to choose the bandwidth N?

Remark 1.22. Thus, we get to the key-point in nonparametric statistics. We have to choose N (or γ or m) in order to balance the bias term and the variance term. As we will see this choice will be difficult since the bias term depends on the unknown f.

1.2.3 Classes of Functions

An important problem now is to define "natural" classes of functions on $\mathscr{F}$.

1.2.3.1 Source Conditions

A standard way to measure the smoothness of the function f is relative to the smoothing properties of the operator A, more precisely in terms of A^*A. Let $\ell :$ $[0,\infty) \to [0,\infty)$ be a continuous, strictly increasing function with $\ell(0) = 0$ and assume that there exists a source $w \in H$ such that

$$f = \ell(A^*A)w, \; w \in H, \; \|w\|^2 \leqslant L. \tag{1.35}$$

This is called a **source condition**. The most standard choice for ℓ is the **Hölder type source condition** where $\ell(x) = x^\mu$, $\mu \geqslant 0$, i.e.

$$f = (A^*A)^\mu w, \; w \in H, \; \|w\|^2 \leqslant L. \tag{1.36}$$

Denote by $\mathscr{F}_\ell(L)$ the class of functions

$$\mathscr{F}_\ell(L) = \left\{ f = \ell(A^*A)w : w \in H, \; \|w\|^2 \leqslant L \right\}. \tag{1.37}$$

In order to take advantage of the source condition we assume that, for any regularization methods, there exists a constant v_0 called **qualification** and a constant $\bar{v}$ such that

$$\sup_{x \in \sigma(A^*A)} |x^v(1 - x\Phi_\gamma(x))| \leqslant \bar{v} \, \gamma^v, \; \forall \, \gamma > 0, \; \forall 0 \leqslant v \leqslant v_0. \tag{1.38}$$

We then get the following theorem in order to control the bias term, i.e. the approximation error.

Theorem 1.4. *Suppose that one has a regularization method $\hat{f}_\gamma$ checking condition (1.38). Define $\mathscr{F}_\ell(L)$ with $\ell(x) = x^\mu$. Then we have*

$$\sup_{f \in \mathscr{F}_\ell(L)} \|\mathbf{E}_f(\hat{f}_\gamma) - f\|^2 \leqslant \bar{v}^2 L \, \gamma^{2\mu},$$

for all $0 \leqslant \mu \leqslant v_0$.

Proof. We have

$$B(\hat{f}_\gamma) = \|\mathbf{E}_f(\hat{f}_\gamma) - f\|^2 = \|\Phi_\gamma(A^*A)A^*Af - f\|^2.$$

Using (1.35), (1.38) and the isometry of the functional calculus we obtain

$$B(\hat{f}_\gamma) = \|(\Phi_\gamma(A^*A)A^*A - I)(A^*A)^\mu w\|^2$$

$$\leqslant L \sup_{x \in \sigma(A^*A)} |x^\mu (1 - x\Phi_\gamma(x))|^2 \leqslant \bar{\nu}^2 L\, \gamma^{2\mu}.$$

The qualification of a method is the largest source condition for which the bias of the method converges with the optimal rate.

For the Landweber iteration, suppose here that $\|A^*A\| = 1$. Note that $\sigma(A^*A) \subset [0,1]$. The approximation error is

$$\sup_{x \in \sigma(A^*A)} |x^\mu (1 - x\Phi_m(x))| \leqslant \sup_{x \in [0,1]} |x^\mu (1 - x)^m|.$$

If we solve this problem, we then obtain that the supremum is attained at point $x_0 = \mu/(\mu + m) \in [0,1]$. Thus, the approximation error is bounded by, for any $\mu > 0$,

$$\sup_{x \in [0,1]} |x^\mu (1 - x)^m| \leqslant \left(\frac{\mu}{\mu + m} \right)^\mu \leqslant Cm^{-\mu}.$$

The qualification of the Landweber method is then ∞, since this result is valid for any $\mu > 0$. Note that $\gamma = 1/m$ here.

The semi-iterative methods are defined via an iteration polynomial Φ_m. The ν-methods have, in fact, been defined such that they minimize

$$\sup_{x \in \sigma(A^*A)} |x^\nu (1 - x\Phi_m(x))|,$$

for all polynomials of degree $m - 1$.

One may prove then (see [49])

$$|1 - x\Phi_m(x)| \leqslant c_\nu (1 + m^2 x)^{-\nu}.$$

Thus, we have

$$\sup_{x \in \sigma(A^*A)} |x^\mu (1 - x\Phi_m(x))| \leqslant c_\nu \sup_{x \in [0,1]} |x^\mu (1 + m^2 x)^{-\nu}|.$$

The maximum is attained at point

$$x_0 = \begin{cases} \mu/(m^2(\nu - \mu)) & \text{if } \mu < \nu, \\ 1 & \text{if } \mu \geqslant \nu. \end{cases}$$

We finally obtain

$$\sup_{x \in \sigma(A^*A)} |x^\mu (1 - x\Phi_m(x))| \leqslant \begin{cases} Cm^{-2\mu} & \text{if } \mu < \nu, \\ Cm^{-2\nu} & \text{if } \mu \geqslant \nu. \end{cases}$$

There is a saturation effect. The qualification of the ν-method is then ν.

Remark 1.23. One very important point here, is that, for the same number of iterations, the approximation error is much better for the ν-method than for the Landweber one. In other terms, one needs m^2 iterations with Landweber and m iterations with v−method for the same accuracy. The Landweber method attains the optimal rates of convergence but with much more iterations than v−method. This is one drawback of the Landweber method in applications.

Some direct computations show that the qualification of the different methods are (see following table).

Table 1.1 Qualification of regularization methods

Method	Qualification
Spectral cut-off	∞
Tikhonov	1
Tikhonov with prior a	$1+2a$
m-iterated Tikhonov	m
Landweber	∞
v-method	v

The main aim now is to understand the precise meaning of source condition on some well-known examples.

1.2.3.2 Ellipsoid of Coefficients

Suppose here that the operator A is compact.

Assuming Hölder type source condition $f = (A^*A)^\mu w$ is then equivalent in the SVD domain to, by functional calculus,

$$f = (A^*A)^\mu w = \sum_{k=1}^{\infty} b_k^{2\mu} w_k \varphi_k,$$

since $w \in H$, where $w_k = \langle w, \varphi_k \rangle$ is in ℓ^2. Denote by $\langle f, \varphi_k \rangle = \theta_k = b_k^{2\mu} w_k$ and since $\|w\|^2 \leqslant L$, we then obtain

$$\|w\|^2 = \sum_{k=1}^{\infty} w_k^2 = \sum_{k=1}^{\infty} b_k^{-4\mu} \theta_k^2 \leqslant L. \tag{1.39}$$

Thus, in the inverse problem framework with compact operator, the source conditions will correspond to the assumption that the coefficients of f belong to some ellipsoid in ℓ^2.

Assume that f belongs to the functional class corresponding to ellipsoids Θ in the space of coefficients $\{\theta_k\}$:

$$\Theta = \Theta(a, L) = \left\{ \theta : \sum_{k=1}^{\infty} a_k^2 \theta_k^2 \leqslant L \right\}, \tag{1.40}$$

where $a = \{a_k\}$ is a non-negative sequence that tends to infinity with k, and $L > 0$. This means that for large values of k the coefficients θ_k will have (a negative) polynomial behaviour in k and will be small.

Remark 1.24. Assumptions on the coefficients θ_k will be usually related to some properties (smoothness) on f. One difficulty in using SVD in inverse problems is that the basis $\{\varphi_k\}$ is defined by the operator A. One then has to hope good properties for the coefficients θ_k of f in this specific basis.

1.2.3.3 Classes of Functions

Suppose that we are exactly in the setting of the periodic convolution of Section 1.1.6.3. Then the operator is compact and the SVD basis is exactly the Fourier basis.

In the special cases where the SVD basis is the Fourier basis, hypothesis on $\{\theta_k\}$ may be precisely written in terms of smoothness for f.

Such classes arise naturally in various inverse problems, they include as special cases the Sobolev classes and classes of analytic functions. In fact, we consider balls of size $L > 0$ in functions spaces.

Let $\{\varphi_k(t)\}$ be the real trigonometric basis on $[0,1]$:

$$\varphi_1(t) \equiv 1, \quad \varphi_{2k}(t) = \sqrt{2}\cos(2\pi kt), \quad \varphi_{2k+1}(t) = \sqrt{2}\sin(2\pi kt), \quad k = 1,2,\ldots.$$

Introduce the **Sobolev classes of functions** (see [12])

$$\mathscr{W}(\alpha,L) = \left\{ f = \sum_{k=1}^{\infty} \theta_k \varphi_k : \theta \in \Theta(\alpha,L) \right\}$$

where $\Theta(\alpha,L)$ with the sequence $a = \{a_k\}$ such that $a_1 = 0$ and

$$a_k = \begin{cases} (k-1)^{\alpha} & \text{for } k \text{ odd,} \\ k^{\alpha} & \text{for } k \text{ even,} \end{cases} \quad k = 2,3,\ldots,$$

where $\alpha > 0, L > 0$.

If α is an integer, this corresponds to the equivalent definition, see Proposition 1.14 in [129],

$$\mathscr{W}(\alpha,L) = \left\{ f \in L^2[0,1] : \int_0^1 (f^{(\alpha)}(t))^2 dt \leqslant L, f^{(j)}(0) = f^{(j)}(1) = 0, j = 0,\ldots,\alpha-1 \right\}$$

where f is 1-periodic and $f^{(\alpha)}$ denotes the weak derivative of f of order α.

In the case where the problem is mildly ill-posed with $b_k = k^{-\beta}$, by (1.39), Hölder type source conditions correspond to

$$\sum_{k=1}^{\infty} b_k^{-4\mu} \theta_k^2 = \sum_{k=1}^{\infty} k^{4\mu\beta} \theta_k^2 \leqslant L,$$

and then $\theta \in \Theta(\alpha, L)$ with $\alpha = 2\mu\beta$.

One may also consider more restrictive conditions and the classes of functions

$$\mathscr{A}(\alpha, L) = \left\{ f = \sum_{k=1}^{\infty} \theta_k \varphi_k : \theta \in \Theta_{\mathscr{A}}(\alpha, L) \right\}$$

where $a_k = \exp(\alpha k)$, $\alpha > 0$, and $L > 0$. This corresponds to the usual classes of **analytical functions**. These functions admit an analytical continuation into a band of the complex plane, see [75]. These functions are thus very smooth (C^{∞}).

In the case where the problem is severely ill-posed with $b_k = \exp(-\beta k)$, by (1.39), Hölder type source conditions correspond to

$$\sum_{k=1}^{\infty} b_k^{-4\mu} \theta_k^2 = \sum_{k=1}^{\infty} e^{4\mu\beta k} \theta_k^2 \leqslant L,$$

and then $\theta \in \Theta_{\mathscr{A}}(\alpha, L)$ with $\alpha = 2\mu\beta$.

1.2.4 Rates of Convergence

1.2.4.1 SVD Setting

In this setting of ill-posed inverse problems with compact operator and functions with coefficients in some ellipsoid, several results have been obtained.

As in Definition 1.9, denote by

$$r_\varepsilon(\Theta) = \inf_{\hat{\theta}} \sup_{\theta \in \Theta} \mathscr{R}(\hat{\theta}, \theta), \tag{1.41}$$

where the $\inf_{\hat{\theta}}$ is taken for all estimators of f, the *minimax risk* on the class of coefficients Θ and the *linear minimax risk*

$$r_\varepsilon^{\ell}(\Theta) = \inf_{\hat{\theta}^{\ell}} \sup_{\theta \in \Theta} \mathscr{R}(\hat{\theta}, \theta),$$

where the $\inf_{\hat{\theta}^{\ell}}$ is among all linear estimators.

There exists a famous result by [109] which exhibits an estimator which is even minimax, i.e. which attains not only the optimal rate, but also the exact constant. This estimator is called the Pinsker estimator.

The following theorem is due to [109].

Theorem 1.5. *Let $\{a_k\}$ be a sequence of non-negative numbers and let $\sigma_k > 0$, $k = 1, 2, \ldots$ Then the linear minimax estimator $\lambda = \{\lambda_k\}$ on $\Theta(a, L)$ is given by*

$$\lambda_k = (1 - c_\varepsilon a_k)_+, \tag{1.42}$$

where c_ε is the solution of the equation

$$\varepsilon^2 \sum_{k=1}^{\infty} \sigma_k^2 a_k (1 - c_\varepsilon a_k)_+ = c_\varepsilon L$$

and the linear minimax risk is

$$r_\varepsilon^\ell(\Theta) = \varepsilon^2 \sum_{k=1}^{\infty} \sigma_k^2 (1 - c_\varepsilon a_k)_+. \tag{1.43}$$

Furthermore, if

$$\frac{\max_{k:a_k<T} \sigma_k^2}{\sum_{k:a_k<T} \sigma_k^2} = o(1), \quad T \to \infty, \tag{1.44}$$

then

$$r_\varepsilon(\Theta) = r_\varepsilon^\ell(\Theta)(1 + o(1)), \tag{1.45}$$

as $\varepsilon \to 0$.

Proof. A proof may be found in [109, 7].

Thus, under the condition (1.44), the linear minimax estimator given by (1.42) is asymptotically minimax among all estimators.

This result has been also generalized to the very specific case of severely ill-posed problems with analytic functions, i.e. when (1.44) is not verified, in [61, 62].

The optimal rates of convergence may also be found, for example in [7] and [32]. The function f is supposed to have Fourier coefficients in some ellipsoid, and the problem is mildly, severely ill-posed or even direct. The rates appear in Table 2.

Example 1.3. All the rates are given for the estimation of a function in one dimension ($d = 1$). Otherwise, in a multidimensional framework, it is well-known that the minimax rates depend on the dimension d.

There exist also many optimal rates results in inverse problems see for example the deconvolution problem in [50, 52], the tomography problem studied in the papers [80, 85, 83, 20], general inverse problems [84, 39, 95, 47, 78, 9] and in related frameworks [19].

A recent review of the different rates in rather general inverse problems may be found in [93].

Table 1.2 Optimal rates of convergence

Inverse Problem/Functions	Sobolev	Analytic
Direct problem	$\varepsilon^{\frac{4\alpha}{2\alpha+1}}$	$\varepsilon^2 \log \frac{1}{\varepsilon}$
Mildly ill-posed	$\varepsilon^{\frac{4\alpha}{2\alpha+2\beta+1}}$	$\varepsilon^2 \left(\log \frac{1}{\varepsilon}\right)^{2\beta+1}$
Severely ill-posed	$\left(\log \frac{1}{\varepsilon}\right)^{-2\alpha}$	$\varepsilon^{\frac{4\alpha}{2\alpha+2\beta}}$

Comments

Ill-posedness. We may remark that the rates usually depend strongly on the smoothness α of the function f and on the degree of ill-posedness β. When β increases the rates are slower. This is a very important point, which characterizes the influence of the ill-posedness in the results. In ill-posed problems the rates are slower, making estimation in these models more difficult.

Direct model/Sobolev. We get the standard rates for nonparametric estimation. Indeed, with the relation $\varepsilon^2 \asymp 1/n$, one really obtains the usual $n^{-\frac{2\alpha}{2\alpha+1}}$ rate for estimating a α smooth function in a nonparametric context, see [74, 122].

The more standard cases for inverse problems are, mildly ill-posed/Sobolev, or severely ill-posed/Analytic. Indeed, they correspond to the natural setting of Hölder source conditions (see Section 1.2.3.2).

Mildly ill-posed/Sobolev. This is, in a way, the more standard framework. One has a not so difficult inverse problem with smooth functions. The rate is then $\varepsilon^{\frac{4\alpha}{2\alpha+2\beta+1}}$. One may see the loss in the rate due to ill-posedness β, compared to the rate in the direct problem $\varepsilon^{\frac{4\alpha}{2\alpha+1}}$. The rate is polynomial in ε and slower than ε^2, as usual in nonparametric statistics.

Severely ill-posed/Analytic. In this context, the problem is very difficult, but the functions are then very smooth. The rate is then still polynomial $\varepsilon^{\frac{4\alpha}{2\alpha+2\beta}}$. This rate is slightly different from the previous case, but related.

The three other cases are very specific problems. The rates are then not polynomial.

Direct model/Analytic. This framework is rather easy. Indeed, the problem is direct, and the functions are very smooth. The rate is then almost parametric, i.e. ε^2. One just looses a logarithmic term compared to the parametric context. From a statistical point of view, the situation is very specific. Indeed, there is no trade-off between bias and variance, the variance term is dominating.

Severely ill-posed/Sobolev. This case corresponds to a very difficult inverse problem with not smooth enough functions. The rate is logarithmic, and thus very slow. From a theoretical point of view, this context might be considered as too difficult. Here, the bias is dominating.

Mildly ill-posed/Analytic. In this case, a mildly ill-posed problem with very smooth functions, the rate is almost the parametric rate ε^2. The variance term is dominating. The functions are so smooth that the inverse problem has almost no influence. Indeed, the degree of ill-posedness appears only in the logarithmic term.

Remark 1.25. One may also consider inverse problems where $\sigma_k \asymp \exp(\beta k^r)$, where $\beta > 0$ and $r \geqslant 1$, for example Heat equation or convolution by a Gaussian kernel. Here the rates will be worse. For example, in the case of Sobolev functions, the rate will be $(\log \frac{1}{\varepsilon})^{-2\alpha/r}$.

Remark 1.26. In the problem of tomography presented in Section 1.1.6.5, the situation is slightly different. Indeed, this is a two dimensional problem. The optimal rate of convergence is given, in [80], and corresponds to $2\alpha/(2\alpha + 3)$. This rate has to be compared to the optimal rate of estimating a function in d dimensions, which is $2\alpha/(2\alpha + d)$. Thus, in dimension $d = 2$, one really sees the ill-posedness $\beta = 1/2$, in obtaining the rate $2\alpha/(2\alpha + 3)$.

In the sequel, we will use quite often the two very standard results,

$$\sum_{k=1}^{n} k^p \approx \frac{n^{p+1}}{p+1}, \; p > -1, \text{ as } n \to \infty \tag{1.46}$$

and

$$\sum_{k=1}^{n} e^{pk} \approx \frac{e^{p(n+1)}}{e^p - 1}, \; p > 0, \text{ as } n \to \infty, \tag{1.47}$$

where $a_n \approx b_n$ means that $a_n/b_n \to 1$ as $n \to \infty$.

In this framework we obtain the following theorem.

Theorem 1.6. *Consider now the case where $\sigma_k \asymp k^\beta, \beta \geqslant 0$ and θ belongs to the ellipsoid $\Theta(\alpha, L)$, where $a_k = k^\alpha, \alpha > 0$. Then the projection estimator with $N \asymp \varepsilon^{-2/(2\alpha+2\beta+1)}$ verifies as $\varepsilon \to 0$*

$$\sup_{\theta \in \Theta(\alpha,L)} R(\theta, N) \leqslant C\varepsilon^{4\alpha/(2\alpha+2\beta+1)}.$$

This rate is optimal (see Theorem 1.5).

Proof. We have,

$$\sup_{\theta \in \Theta(\alpha,L)} R(\theta, N) = \sup_{\theta \in \Theta(\alpha,L)} \sum_{k=N+1}^{\infty} \theta_k^2 + \varepsilon^2 \sum_{k=1}^{N} \sigma_k^2.$$

We bound the first term as follows,

$$\sup_{\theta \in \Theta(\alpha,L)} \sum_{k=N+1}^{\infty} \theta_k^2 \leqslant \sup_{\theta \in \Theta(\alpha,L)} \sum_{k=N+1}^{\infty} k^{2\alpha} \theta_k^2 k^{-2\alpha}$$

$$\leqslant N^{-2\alpha} \sup_{\theta \in \Theta(\alpha,L)} \sum_{k=1}^{\infty} k^{2\alpha} \theta_k^2 \leqslant L N^{-2\alpha}.$$

The variance term is controlled by

$$\varepsilon^2 \sum_{k=1}^{N} \sigma_k^2 \asymp \varepsilon^2 \sum_{k=1}^{N} k^{2\beta} \asymp \frac{\varepsilon^2 N^{2\beta+1}}{2\beta + 1},$$

when N is large, by use of (1.46). Thus,

$$\sup_{\theta \in \Theta(\alpha,L)} R(\theta,N) \leqslant L\,N^{-2\alpha} + \frac{\varepsilon^2 N^{2\beta+1}}{2\beta+1}.$$

If we want to attain the optimal rate of convergence we have to choose N of order $\varepsilon^{-2/(2\alpha+2\beta+1)}$ as $\varepsilon \to 0$. This choice corresponds to the trade-off between the bias term and the variance term.

Remark 1.27. This proof is very simple and only concerns the rate of convergence for a given estimator, the so-called upper bound. The proof of an upper bound for some estimator is usually rather easy. There is no proof here of the lower bound, i.e. showing that no estimator has a risk converging faster. The lower bound is proved by Theorem 1.5. Nevertheless, lower bounds are very important in nonparametric statistics. Indeed, it is the lower bound which proves that the estimator is optimal, i.e. one of the best estimator in a given model. For a discussion in details of the standard methods, see [129].

Remark 1.28. Considering the minimax point of view, we may remark that there exists an optimal choice for N which corresponds to the balance between the bias and the variance. However, this choice depends very precisely on the smoothness α and on the degree of ill-posedness of the inverse problem β.

Even in the case where the operator A (and then its degree β) is known, it has no real meaning to consider that we know the smoothness of the unknown function f.

These remarks lead to the notion of adaptation and also oracle inequalities, i.e. how to choose the bandwidth N without strong a priori assumptions on f (see Section 1.3).

1.2.4.2 Deconvolution on $\mathbb{R}$

Assume that we are in the special inverse problem of deconvolution on $\mathbb{R}$ (see Section 1.1.7.2). Consider only the case of spectral cut-off regularization. We estimate in the Fourier domain Ff by

$$\frac{FY(\omega)}{\tilde{r}(\omega)}I(\omega : \tilde{r}^2(\omega) > \gamma),$$

and then the spectral cut-off regularization is

$$\hat{f}_\gamma^{SC} = F^{-1}\left(\frac{FY(\omega)}{\tilde{r}(\omega)}I(\omega : \tilde{r}^2(\omega) > \gamma)\right).$$

As in the SVD case, the bias term (approximation error) is usually controlled by the source conditions. In this framework, the Hölder source condition (1.36) is equivalent to, in the Fourier domain,

$$Ff = \tilde{r}^{2\mu}Fw, \; w \in L^2(\mathbb{R}), \; \|w\|^2 \leqslant L.$$

$$\int_{\mathbb{R}} |Fw(\omega)|^2 d\omega = \int_{\mathbb{R}} |Ff(\omega)|^2 \tilde{r}^{-4\mu}(\omega) d\omega \leqslant L. \qquad (1.48)$$

Similarly to the previous section, if $\tilde{r}(\omega) = |\omega|^{-\beta}$, the problem is then mildly ill-posed. In this case, the source conditions correspond to some Sobolev class of functions on $\mathbb{R}$ (see [10])

$$\mathcal{W}(\alpha, L) = \left\{ f \in L^2(\mathbb{R}) : \int_{\mathbb{R}} |\omega|^2 |Ff(\omega)|^2 \leqslant L \right\},$$

which is equivalent to, for $\alpha \in \mathbb{N}$,

$$\mathcal{W}(\alpha, L) = \left\{ f \in L^2(\mathbb{R}) : \int_{\mathbb{R}} (f^{(\alpha)}(t))^2 dt \leqslant L \right\}.$$

We then obtain

$$(\mathbf{E}_f \hat{f}_\gamma^{SC}(x) - f(x)) = \frac{1}{\sqrt{2\pi}} \int_{\mathbb{R}} e^{-i\omega x} \left(Ff(\omega)I(\omega : \tilde{r}^2(\omega) > \gamma)) - Ff(\omega) \right) d\omega,$$

and then for the bias

$$\int_{\mathbb{R}} (\mathbf{E}_f \hat{f}_\gamma^{SC}(x) - f(x))^2 dx \leqslant \frac{1}{2\pi} \int_{|\omega| > \gamma^{-1/2\beta}} |Ff(\omega)|^2 d\omega$$

$$\leqslant \gamma^{4\mu\beta/2\beta} \int_{\mathbb{R}} |Ff(\omega)|^2 |\omega|^{4\mu\beta} d\omega \leqslant L\gamma^{2\mu}.$$

We need now to bound the stochastic term. Using (1.32), we have

$$\mathbf{E}_f \|\hat{f}_\gamma^{SC} - \mathbf{E}_f(\hat{f}_\gamma^{SC})\|^2 \leqslant \mathbf{E}\|\varepsilon \Phi_\gamma(A^*A)A^*\xi\|^2.$$

Using (1.22) and Lemma 1.5, we may bound the variance term

$$\mathbf{E}_f \|\hat{f}_\gamma^{SC} - \mathbf{E}_f(\hat{f}_\gamma^{SC})\|^2 \leqslant \frac{\varepsilon^2}{2\pi} \mathbf{E} \int_{|\omega| < \gamma^{-1/2\beta}} \left| \frac{\eta(\omega)}{\tilde{r}(\omega)} \right|^2 d\omega$$

$$\leqslant \frac{\varepsilon^2}{2\pi} \int_{|\omega| < \gamma^{-1/2\beta}} |\omega|^{2\beta} d\omega \asymp \varepsilon^2 \, (\gamma^{-1/2\beta})^{2\beta+1} = \varepsilon^2 \, \gamma^{-\frac{2\beta+1}{2\beta}}.$$

The risk of the spectral cut-off is then bounded by

$$\mathbf{E}_f \|\hat{f}_\gamma^{SC} - f\|^2 \leqslant L\gamma^{2\mu} + C\gamma^{-\frac{2\beta+1}{2\beta}},$$

the optimal choice is then $\gamma^* \asymp (\varepsilon^2)^{2\beta/(4\mu\beta+1)}$ which corresponds to the rate

$$\mathbf{E}_f \|\hat{f}_{\gamma^*}^{SC} - f\|^2 \leqslant C(\varepsilon^2)^{\frac{4\mu\beta}{4\mu\beta+2\beta+1}}.$$

This rate may be shown to be optimal.

Remark 1.29. Recall that here $N = [\gamma^{-1/2\beta}]$. The rates are in fact the same than in the compact case of Section 1.2.4.1.

In the case of severely ill-posed problems, i.e. $\tilde{r}(\omega) = \exp(-\beta|\omega|)$, the class coming from Hölder source condition is then different. By using (1.48), we have

$$\int_{\mathbb{R}} |Fw(\omega)|^2 d\omega = \int_{\mathbb{R}} |Ff(\omega)|^2 \tilde{r}^{-4\mu}(\omega) d\omega \leqslant L.$$

Thus,

$$\int_{\mathbb{R}} |Ff(\omega)|^2 \exp(4\mu\beta|\omega|) d\omega \leqslant L.$$

which corresponds to the the class of analytic functions, i.e. which admits an analytic continuation into a band of the complex plane, see for example [75].

Remark 1.30. In the context of general inverse problems, with general regularization methods, it is also possible to obtain results concerning rates of convergence (see [9]).

1.2.5 Comparison Between Deterministic and Stochastic Noise

In this section, consider the model of inverse problems with deterministic noise. This model is, in some sense, the historical model of inverse problems. It appears for example in [127] and [128]. The analog of the stochastic model (1.2), in the deterministic framework is the following. We have

$$Y = Af + \varepsilon h, \tag{1.49}$$

where the noise h is considered as some deterministic element $h \in G$, with $\|h\| \leqslant 1$. Since the noise is some unknown element of a ball in G, then the results have to be obtained for any possible noise, i.e. for the worst noise.

Compare the deterministic model in (1.49) and the stochastic model in (1.2) where ξ is a white noise. At first glance, it may seem, that the main difference between the two models concerns the nature of the noise, deterministic against stochastic.

In fact, it is more the level of the two noises which are not the same.

The first main difference, since ξ is a white noise, is that Y in (1.2) is not really "observed". Indeed, ξ does not take its values in G. We only observe its projection on some basis. Indeed, ξ as a white noise, is not a Hilbert-space random variable in G but a Hilbert-space process acting on G. Formally, we have $\|\xi\|_G = \infty$, thus ξ is not an element of G. On the other hand, the deterministic noise h belongs to G, and $\|h\| \leqslant 1$. The deterministic noise is then "small" compared to the stochastic one.

This fact, has been already noted in [38].

In order to have a more comprehensive study, consider the class of linear injective and compact operators which admit a singular value decomposition (SVD) (see Section 1.1.5).

The analog of the sequence space model in (1.6) may be written as

$$X_k = \theta_k + \varepsilon \sigma_k \, h_k, \quad k = 1, 2, \ldots, \tag{1.50}$$

where $\{h_k\}$ are the coefficients of h in the basis $\{\psi_k\}$.

A natural way of studying the two frameworks is to compare the accuracy of estimation (reconstruction). Define two standard criteria, in order to measure the error or risk, for any estimator $\hat{f}$ (or regularization method). For the stochastic noise model, use the maximal risk defined in Definition 1.9. For the deterministic noise model, define the worst noise risk

$$\sup_{f \in \mathcal{F}} \sup_{\|h\| \leqslant 1} \|\hat{f} - f\|^2,$$

where f belongs to some class of functions $\mathcal{F}$.

The goal is to compare the optimal rates of convergence in each model, i.e. the order of the risk of the best possible estimator as $\varepsilon \to 0$. Indeed, this rate defines a notion of difficulty of estimation in a given model. Two models with the same optimal rates of convergence are usually thought to be close, at least from the estimation point of view.

One difference between deterministic and stochastic cases, is that since $\|h\| \leqslant 1$ (i.e. $\sum h_k^2 \leqslant 1$), the noise h_k decreases in (1.50) as k increases. In the stochastic case, the level of the noise ξ_k is the same in each coefficient X_k. Thus, the stochastic noise seems to be larger.

It is well-known, that the rates of convergence depend on difficulty of the inverse problem and smoothness conditions on the function f (see Section 1.2.4). For the inverse problems, the two standard cases are $\sigma_k \asymp k^\beta$ or $\sigma_k \asymp e^{\beta k}$, $\beta > 0$ which correspond to mildly or severely ill-posed respectively. The parameter β denotes the degree of ill-posedness.

Concerning smoothness properties of f, associated with the behaviour of its coefficients θ_k, consider the ellipsoid of coefficients in ℓ^2 as in Section 1.2.3.2. Consider the two standard cases, Sobolev ($a_k = k^\alpha$) and Analytic ($a_k = e^{\alpha k}$), where $\alpha > 0$ is the smoothness of f.

For the stochastic noise, the optimal rates of convergence may be found in Table 2. Concerning the deterministic noise, rates of convergence may be obtained, for example in [49].

Consider here the two more natural cases, *polynomial* ($\sigma_k \asymp k^\beta$ and $a_k = k^\alpha$) and *exponential* ($\sigma_k \asymp e^{\beta k}$ and $a_k = e^{\alpha k}$).

Consider also a third case: the direct problem, where $\sigma_k \equiv 1$ (i.e. $A = I$) and f belongs to a Sobolev ball ($a_k = k^\alpha$).

All these rates are given in the following table:

Remark that in the exponential case, rates of convergence are the same for the two models.

Table 1.3 Rates for deterministic and stochastic model

	Deterministic	Stochastic
Direct	ε^2	$\varepsilon^{\frac{4\alpha}{2\alpha+1}}$
Polynomial	$\varepsilon^{\frac{4\alpha}{2\alpha+2\beta}}$	$\varepsilon^{\frac{4\alpha}{2\alpha+2\beta+1}}$
Exponential	$\varepsilon^{\frac{4\alpha}{2\alpha+2\beta}}$	$\varepsilon^{\frac{4\alpha}{2\alpha+2\beta}}$

On the other hand, rates are different in the polynomial case, which is more standard. There is a small difference between $2\alpha/(2\alpha+2\beta)$ and $2\alpha/(2\alpha+2\beta+1)$ which could be thought as not very important. However, this is fundamental.

In order to understand well this phenomenon, consider what happens when $\beta \to 0$. The problem is less and less ill-posed and becomes close to the case $\beta = 0$, i.e. to the direct case where $\sigma_k \equiv 1$ and $A = I$ is the identity. In the deterministic problem, the rate will attain ε^2 $(a = 1)$ in the direct case. In the stochastic framework, the rate will be $\varepsilon^{4\alpha/(2\alpha+1)}$.

The fundamental difference now appears. In the stochastic direct problem, the rate depends on the smoothness α of the estimated function f. This is not true for the deterministic framework.

In the stochastic case, in order to estimate the function f, one needs to balance the approximation error and the stochastic error. This is the usual trade-off in non-parametric statistics between the bias and the variance.

Everything is different in the deterministic case. The function f will be directly estimated by Y, which attains the rate ε^2. There is no trade-off, the whole series $\{X_k\}$ is used to estimate $\{\theta_k\}$. The rate ε^2 is usually obtained in statistics in the parametric case, i.e. when estimating a vector θ of finite dimension. In the stochastic case, one cannot use directly Y which has infinite risk.

In the direct case, the two models are thus totally different. Indeed, the deterministic noise is smaller than the stochastic one, because it is bounded. In (1.50) the errors h_k become small with k, whereas the stochastic errors ξ_k are of the same order in (1.6).

From a statistical point of view such a small error would not really make sense. Indeed, statistics study the effect of stochastic errors and these errors should be important enough. However, from a numerical point of view, it could make sense to neglect the noise, or at least to consider it as small. Thus, the difference is more in the level of the noise than its nature (deterministic or stochastic).

In order to explain more clearly the influence of noise, consider the simple projection estimator,

$$\hat{\theta}_k = I(k \leqslant N) \frac{y_k}{b_k},$$

where $I(\cdot)$ is the indicator function and N is some integer. It is known that this family of estimators attains, for a correct choice of N, the optimal rate of convergence on Θ (see Theorem 1.6).

The ℓ^2−risk of this estimator is, in the stochastic model,

$$\mathbf{E}_\theta \|\hat\theta - \theta\|^2 = \mathbf{E}_\theta \sum_{k=1}^\infty (\hat\theta_k - \theta_k)^2 = \sum_{k=N+1}^\infty \theta_k^2 + \varepsilon^2 \sum_{k=1}^N \sigma_k^2, \tag{1.51}$$

and the ℓ^2-error of the reconstruction method, in the deterministic model, is

$$\sup_{\|h\|\leqslant 1} \|\hat\theta - \theta\|^2 = \sup_{\|h\|\leqslant 1} \left(\sum_{k=N+1}^\infty \theta_k^2 + \varepsilon^2 \sum_{k=1}^N h_k^2 \sigma_k^2 \right) = \sum_{k=N+1}^\infty \theta_k^2 + \varepsilon^2 \sigma_N^2, \tag{1.52}$$

in the case of increasing σ_k.

The influence of the inverse problem is only on the variance term, i.e. the second term in the right-hand side of (1.51) and (1.52). The approximation error $\sum_{k>N} \theta_k^2$ is not modified by the ill-posedness of the inverse problem.

The following table gives the order, as $\varepsilon \to 0$, of the variance term $\varepsilon^2 \sum_{k=1}^N \sigma_k^2$ or $\varepsilon^2 \sigma_N^2$, in the various settings.

Table 1.4 Variances for deterministic and stochastic model

	Deterministic	Stochastic
Direct	ε^2	$\varepsilon^2 N$
Mildly	$\varepsilon^2 N^{2\beta}$	$\varepsilon^2 N^{2\beta+1}$
Severely	$\varepsilon^2 e^{\beta N}$	$\varepsilon^2 e^{\beta N}$

The direct case corresponds to $b_k \equiv 1$. In the deterministic model, since $h \in \ell^2$, the variance term is ε^2, and does not depend on N. Thus, there is no trade-off, N can be chosen as ∞, or any choice such that $\sum_{k>N} \theta_k^2 = O(\varepsilon^2)$.

In the stochastic case, the variance term is $\varepsilon^2 N$. Thus, we have to balance the bias and the variance, as usually in nonparametric statistics, and find the optimal choice of N.

The variance terms stay different in the mildly ill-posed (polynomial) case. The ratio between the two variance terms is again N. However, this difference is less important as β increases. Indeed, when β is large $N^{2\beta+1}$ is close to $N^{2\beta}$.

The main point is that the variance term is larger in the case of ill-posed problems. The degree of ill-posedness β appears directly in the variance term. The variance increases with β.

Thus, in the case of ill-posed inverse problems, the deterministic error has more influence than for the direct case. The presence of β increases the variance term.

For large β, the two models give almost the same rates. Finally, for severely ill-posed problems (exponential case), these rates are the same.

The ill-posedness of the inverse problem hides, in some sense, the difference between the two kinds of noise, by increasing the small deterministic noise. When β is large, the main part of the variance term $\varepsilon^2 N^{2\beta(+1)}$ is due to the inverse problem and not to the nature of the noise. The inverse problem makes these two models more close, when for the direct problem they are completely different.

The difference between deterministic and stochastic noise is in its level and not really in its nature. Thus, a stochastic model with a small noise could be considered. The model is the following,

$$X_k = \theta_k + \varepsilon \; \sigma_k e_k \xi_k, \quad k = 1, 2, \ldots, \tag{1.53}$$

where $\{\xi_k\}$ are independent standard Gaussian random variables, and $\{e_k\} \in \ell_2$, $\|e\| = 1$. The risk of a projection estimator is then

$$\mathbf{E}_\theta \|\hat\theta - \theta\|^2 = \sum_{k=N+1}^{\infty} \theta_k^2 + \varepsilon^2 \sum_{k=1}^{N} e_k^2 \sigma_k^2. \tag{1.54}$$

In the direct case, the variance term is then $\varepsilon^2 \sum_{k=1}^{N} e_k^2$, bounded by ε^2. The optimal rate is so ε^2, as for the deterministic case.

In the case of ill-posed problem, hypothesis should be more precise in order to obtain explicit rate of convergence. Indeed, in the deterministic case we study the worst noise, i.e. $\sup_{\|h\| \leqslant 1}$. Thus, we have to consider a noise in ℓ^2 but rather large, almost on the "edge". Some example is $e_k = (\sqrt{k} \log(k+1))^{-1}$, which is in ℓ^2. It is clear that dividing X_k by e_k, one obtains a model equivalent to (1.53), with a new $\sigma_k' = \sigma_k e_k$.

With this choice of $\{e_k\}$, in the mildly ill-posed case, the variance term is then $\varepsilon^2 \sum_{k=1}^{N} k^{2\beta-1} \log^{-1}(k+1)$, which is equivalent (up to a log term) to $\varepsilon^2 N^{2\beta}$. Thus, the risk in (1.54) is of the same order than in the deterministic case. Looking at the rate for the stochastic case with $\beta - 1/2$, we obtain the rate with β for deterministic case.

In the exponential case, $\{e_k\}$ has no real influence.

Thus, using a model of inverse problem with stochastic noise with a "small" noise, we obtain the same rate of convergence than for the deterministic case (up sometimes to some log term). A "small" stochastic noise is in fact a Hilbert-space random variable, and not only a Hilbert-space process. It is random, but really takes its values in G.

In conclusion, the main difference between the two approaches comes more from the level of the noise and not so much from its nature.

However, this short study is not at all exhaustive. A more precise approach, based not only on the comparison between the optimal rates, but also the exact constants in the risk, would highlight more differences. For all that, such a technical comparison would not really make sense, since at this precision level, any models are different.

A more sensible framework, in order to compare deterministic and stochastic noise, concerns the construction of adaptive estimators, i.e. which do not depend on the smoothness α of the function to reconstruct (see the following Section 1.3).

In this case the nature of the noise would have more influence. Indeed, the methods could then be very different, for example the discrepancy principle [49] for deterministic noise, or cross-validation, unbiased risk estimator (see Section 1.3.3.1) or the Lepski method [87, 101] for stochastic noise. In the deterministic case, one crucial point is that the error in the data (ε) is precisely known, and then, one can reject reconstruction $\hat{f}$ such that $\|A\hat{f} - Y\| > \varepsilon$. In the stochastic case, the main idea of adaptation is to use large deviations for the noise. Usually, one find values such that the noise will have a very small probability to fall beyond, as in Lemma 1.7 (see also for some examples of adaptivity results [78] and [25]).

In conclusion, this study is not claiming that the two approaches present no difference. The two frameworks are similar in some ways. The differences coming more from the level of the noise than from its nature.

1.3 Adaptation and Oracle Inequalities

One of the most important point in nonparametric statistics is then typically linked to the problem of calibrating by the data the tuning parameter (N, γ or m) in any class of estimators. For example, we have seen that this choice is very sensitive if we want to attain the optimal rate of convergence.

This problem leads to the notion of adaptation and oracle inequalities, i.e. how to construct truly data-driven estimators which have good theoretical properties.

This framework is very important, in theory, but also in applications. Indeed, the notion of, rates of convergence, smoothness α of the function to reconstruct, degree of ill-posedness β of the inverse problem, are very interesting. They help to understand, the difficulty of an ill-posed problem, the influence of smoothness on the rates and so on... The (minimax) optimality of an estimator is also very important. Indeed it shows that no estimator may do better in a given class of functions.

However, they are just mathematical and asymptotical tools. The degree β of a given inverse problem is usually not known. It is even worse concerning the smoothness α of the target function f. One has no chance to have any idea of it.

Definitely, one cannot rely on some unknown smoothness, and asymptotic relationship, in order to make the choice of the tuning parameter (N or γ). One has to really construct data-driven methods in order to calibrate the tuning parameter. Then, the main goal is to prove that this data-driven method has a good behaviour, from a mathematical point of view.

This problem of adaptation is presented in the framework of the sequence space model defined in (1.6) and directly linked, by use of the SVD, to some inverse problem with a compact operator A.

1.3.1 Minimax Adaptive Procedures

The starting point of the approach of **minimax adaptation** is a collection $\mathcal{G} = \{\Theta_\alpha\}$ of classes $\Theta_\alpha \subset \ell^2$. The statistician knows that θ belongs to some member Θ_α of the collection $\mathcal{G}$, but he does not know exactly which one. If Θ_α is a smoothness class, this assumption can be interpreted as follows: the statistician knows that the underlying function has some smoothness, but he does not know the degree of smoothness.

Definition 1.12. An estimator θ^* is called **minimax adaptive** on the scale of classes $\mathscr{G}$ if for every $\Theta_\alpha \in \mathscr{G}$ the estimator θ^* attains the optimal rate of convergence.

An estimator θ^* is called **sharp minimax adaptive** on the scale of classes $\mathscr{G}$ if it also attains the exact minimax constant.

The idea of choosing the tuning parameter (bandwidth) of an estimator in a data-driven way is a very standard idea in nonparametric statistics. However, the main difficulty then concerns the mathematical behaviour of such an estimator. Only quite recently, this idea has been formalized in a rigorous way by [87].

Lepski, in [88], has developed a method in order to construct adaptive estimators, i.e. an estimator which attains the optimal rate for any class Θ_α.

In some cases, no estimator attains (exactly) the optimal rate on the whole scale. One has often to pay a price for adaptation [89]. This cost in the accuracy for construction of adaptive estimator is usually the loss of a logarithmic term in the rate of convergence.

Since the beginning of the 90's, adaptive estimation is really one of the leading topics in nonparametric statistics. Many adaptive (or almost adaptive) estimators have been constructed, in very different frameworks, and various classes of functions.

One may use very different procedures in order to construct adaptive estimators, for example, Lepski's algorithm in [88], model selection in [4], unbiased risk estimation in [82], or wavelets thresholding in [40].

Adaptive minimax estimation in statistical inverse problems as (1.2) has been studied quite recently. This has been done for many inverse problems (deconvolution, heat equation, tomography...).

There exist also a very vast literature on adaptation in inverse problems by Wavelet-Vaguelette Decomposition (WVD) on the Besov scale of classes, see [39, 83, 78, 28, 34, 79, 29, 71].

Lepski's procedure has been also used in inverse problems in several papers [55, 56, 21, 5, 101].

The unbiased risk estimation is also quite popular in inverse problems, see [25, 97].

The model selection is considered in inverse problems [35, 91].

Other adaptive results may be found in [46, 47, 48, 58, 24, 57].

Remark 1.31. Minimax adaptive estimators are really important in statistics from a theoretical and from a practical point of view. Indeed, it implies that these estimators are optimal for any possible parameter in the collection $\mathscr{G}$. From a more practical point of view it garantees a good accuracy of the estimator for a very large choice of functions.

Thus, we have an estimator which automatically adapts to the unknown smoothness of the underlying function. The estimator is then completely data-driven and automatic. However, it behaves as if it knew the true smoothness. This notion is very important since this smoothness is almost never known.

1.3.2 Oracle Inequalities

Consider now a linked, but different point of view. Assume that a class of estimators is fixed, i.e. that the class of possible weights Λ is given. Define the **oracle** λ^0 as

$$R(\theta, \lambda^0) = \inf_{\lambda \in \Lambda} R(\theta, \lambda). \qquad (1.55)$$

The oracle corresponds to the best possible choice in Λ, i.e. the one which minimizes the risk. However, this is not an estimator since the risk depends on θ, the oracle will also depend on this unknown θ. For this reason, it is called oracle since it is the best one in the family, but it knows the true θ. Another important point is to note that the oracle λ^0 usually depends really on the family Λ. As an infimum, the oracle is not necessarily unique or may not be exactly attained. However, this has no influence on the results. Indeed, one only considers the risk of the oracle $\inf_{\lambda \in \Lambda} R(\theta, \lambda)$.

The goal is then to find a data-driven sequence of weights $\lambda^\star$ with values in Λ such that the estimator $\theta^\star = \hat{\theta}(\lambda^\star)$ satisfies an **oracle inequality**, for any $\varepsilon > 0$ and any $\theta \in \ell^2$, there exits $\tau_\varepsilon > 0$,

$$\mathbf{E}_\theta \| \theta^\star - \theta \|^2 \leqslant (1 + \tau_\varepsilon) \inf_{\lambda \in \Lambda} R(\theta, \lambda) + \Omega_\varepsilon, \qquad (1.56)$$

where Ω_ε is some positive remainder term and $\tau_\varepsilon > 0$ (close to 0 if possible). If the remainder term is small, i.e. smaller than the main term $R(\theta, \lambda^0)$ then an oracle inequality proves that the estimator has a risk of the order of the oracle.

A standard remainder term is $\Omega_\varepsilon = c\varepsilon^2$, where c is uniform positive constant. In this case, the remainder term is really considered as "small". Indeed, in most of the nonparametric frameworks, the rates of convergence are worse than ε^2, which is the parametric rate (see Table 2). In an asymptotic point of view, the risk of the oracle, will then be larger than the remainder term. Thus, the leading term of the inequality will be the risk of the oracle.

A more precise result is the following. The estimator $\theta^\star = \hat{\theta}(\lambda^\star)$ satisfies an **exact oracle inequality**, for any $\varepsilon > 0$, any $\theta \in \ell^2$, and for all $\tau_\varepsilon > 0$,

$$\mathbf{E}_\theta \| \theta^\star - \theta \|^2 \leqslant (1 + \tau_\varepsilon) \inf_{\lambda \in \Lambda} R(\theta, \lambda) + \Omega_\varepsilon, \qquad (1.57)$$

where $\Omega_\varepsilon \geqslant 0$ and usually Ω_ε depends on τ_ε.

Remark 1.32. We are interested in data-driven methods, and thus automatic, which more or less mimic the oracle.

One may obtain some asymptotic results when $\varepsilon \to 0$. We call an **asymptotic exact oracle inequality** on the class Λ, as $\varepsilon \to 0$,

$$\mathbf{E}_\theta \| \theta^\star - \theta \|^2 \leqslant (1 + o(1)) \inf_{\lambda \in \Lambda} R(\theta, \lambda), \qquad (1.58)$$

for every θ within some large subset $\Theta_0 \subseteq \ell^2$.

In other words, the estimator θ^* asymptotically precisely mimics the oracle on Λ for any sequence $\theta \in \Theta_0$.

An important question is how large is the class Θ_0 for which (1.58) can be guaranteed. Ideally, we would like to have (1.58) for all $\theta \in \ell^2$ and with $o(1)$ that is uniform over $\theta \in \ell^2$ (i.e. $\Theta_0 = \ell^2$). This property can be obtained for some classes Λ (see, for example, [25]), but with restrictions on Λ that do not allow correct rates of the oracle risk $R(\theta, \lambda^0)$ for "very smooth" θ, i.e. analytic functions. If we choose Λ large enough to allow all the spectrum of rates for the oracle risk, up to the parametric rate ε^2, we cannot have (1.58) for all $\theta \in \ell^2$ and with $o(1)$ that is uniform over $\theta \in \ell^2$. Although, slightly restricted versions of (1.58) are possible. In particular, Θ_0 can be either the set of all $\theta \neq 0$, or the set ℓ^2_-, i.e. the subspace of ℓ^2 containing all the sequences with infinitely many non-zero coefficients (i.e. "nonparametric" sequences), or the set $\{\theta : \|\theta\| \geq r_0\}$ for some small $r_0 > 0$. Also, the uniformity of $o(1)$ in θ is not always granted if both classes Λ and Θ_0 are large.

One of first to really see the importance of oracle inequalities are Donoho and Johnstone in [40] where they introduced also the name *oracle*.

During the end of 90's, the oracle was still mainly seen as just a tool in order to prove adaptation. However, nowadays, this point of view has really changed. Oracle inequalities are often considered as the main results for a given estimator. The oracle approach has also modified the statisticians behaviour. For example, non-asymptotical point of view is much more common now.

To our knowledge, one of the first exact oracle inequalities were obtained for the classes of "ordered linear smoothers" in [82]. In particular, Kneip's result applies to projection estimators and to spline smoothing.

The work of Birgé and Massart on model selection is also strongly related to the notion of oracle inequalities, usually in a slightly different form with a penalized version of an oracle inequality, see [4, 8, 100].

Oracle inequalities are nowadays popular, in the nonparametric statistics literature, see [41, 17, 105, 31, 114].

The earlier papers of [119, 90, 59, 111] also contain, although implicitly, oracle inequalities for some classes Λ. All these papers use the unbiased risk estimation method (see Section 1.3.3.1).

A very interesting review on the topic is [18].

The oracle approach is quite recent in inverse problems. However, the oracle point of view, was growing at the same times than the statistical study of inverse problems. Thus, there is now a rather large interest on oracle inequalities in the statistical inverse problem community, see [78, 25, 60, 35, 91, 98].

Comments

Oracle/minimax. The oracle approach is in some sense the opposite of the minimax approach. Here, we fix a family of estimators and choose the best one among them. In the minimax approach, on the other hand, one tries to get the best accuracy for functions which belong to some function class. The oracle approach is really based on classes of estimators, when the minimax approach is built on classes of functions.

Non-asymptotic oracle. The oracle inequalities, are true for any θ, and are non asymptotic. This fact has really changed the point of view concerning nonparametric statistics. Nowadays, non-asymptotic results are really popular.

Oracle: tool for adaptation. The oracle approach is often used as a tool in order to obtain adaptive estimators. Indeed, the oracle in a given class often attains the optimal rate of convergence. Moreover, the estimator does not depend on any smoothness assumptions on f. Thus, by proving an oracle inequality, one often obtains, a minimax adaptive estimator, see for example, Theorems 1.8 and 1.10. During a quite long time, oracle inequalities were mainly considered as just a tool in order to get minimax adaptive results. Already, [40] pointed out that minimax adaptation can be proved as a consequence of oracle inequalities. They also showed that the method of Stein's unbiased risk estimator is minimax sharp adaptive (or almost minimax sharp adaptive) on some Besov classes.

Minimax: justification for oracle. On the other hand, nowadays, the minimax theory may be viewed as a justification for oracle inequality. Indeed, one may ask if the given family of estimators is satisfying. One possible mathematical answer comes from minimax results, which prove that a given family gives optimal estimators. However, in applications, scientists are usually convinced that their favourite method (Tikhonov, projection, ν−method,...) is satisfying.

1.3.3 Model Selection

Usually, one key assumption in this approach of oracle inequality, is that λ^* is restricted to take its values in the same class Λ that appears in the RHS of (1.56). A **model selection** interpretation of (1.56) is the following: in a given class of models Λ we pick the model λ^* that is the closest to the true parameter θ in terms of the risk $R(\theta, \lambda)$.

The framework of model selection is very popular in statistics, and may have several meaning depending on the topics. We consider the model selection approach to the problem of choosing, among a given family of models Λ (estimators), the best possible one. This choice should be made based on the data and not due to some a priori information on the unknown function f.

1.3.3.1 Unbiased Risk Estimation

The definition of the oracle in (1.55) is that it minimizes the risk. Since θ is unknown, the risk is also, and so is the oracle.

A very natural idea in statistics is to estimate this unknown risk by a function of the observations, and then to minimize this estimator of the risk. A classical approach to this minimization problem is based on the principle of **Unbiased Risk Estimation** (URE).

The idea of unbiased risk estimation was developed in [2] and also in [96, 121]. This problem was originally studied in the framework of parametric estimation where the dimension of the model had to choosen.

Mallows, in [96], introduced the C_p in the specific context of regression and the problem of selecting the number of variables that one wants to use in the model.

Akaike, in [2], proposed the Akaike Information Criteria (AIC) in a rather general setting. The idea is to choose the number of parameters N in order to minimize $-2\mathscr{L}_N + 2N$ where $\mathscr{L}_N$ is the maximal value of the log-likelihood, see [3]. In our framework of Gaussian white noise and sequence space model, i.e. a Gaussian noise with a known variance, then AIC and C_p are equivalent. There exist now a very large number of criteria many of them related to AIC, see [119, 90, 111] or the Bayesian Information Criteria (BIC) in [118].

Stein, in [121], proposed his well-known version of URE as the *Stein Unbiased Risk Estimation* (SURE). The results are specific to the Gaussian framework.

Nowadays, C_p, AIC, BIC are all used as basic data-driven choices for many statistical models and in several standard softwares.

This idea appears also in all the cross-validation techniques, see the Generalized Cross-Validation (GCV) in [36].

In inverse problems, the URE method is studied in [25], where exact oracle inequalities for the mean square risk were obtained.

In this setting, the functional

$$\mathscr{U}(X,\lambda) = \sum_{k=1}^{\infty}(1-\lambda_k)^2(X_k^2 - \varepsilon^2\sigma_k^2) + \varepsilon^2\sum_{k=1}^{\infty}\sigma_k^2\lambda_k^2 \qquad (1.59)$$

is an unbiased estimator of $R(\theta,\lambda)$ defined in (1.34).

$$R(\theta,\lambda) = \mathbf{E}_\theta\mathscr{U}(X,\lambda), \ \forall\lambda. \qquad (1.60)$$

The principle of unbiased risk estimation suggests to minimize over $\lambda \in \Lambda$ the functional $\mathscr{U}(X,\lambda)$ in place of $R(\theta,\lambda)$. This leads to the following data-driven choice of λ:

$$\lambda_{ure}^\star = \arg\min_{\lambda\in\Lambda}\mathscr{U}(X,\lambda) \qquad (1.61)$$

and the estimator $\theta_{ure}^\star$ defined by

$$\theta_k^\star = \lambda_k^\star X_k. \qquad (1.62)$$

Let the following assumptions hold. For any $\lambda \in \Lambda$

$$\textbf{(A1)} \quad 0 < \sum_{k=1}^{\infty}\sigma_k^2\lambda_k^2 < \infty, \quad \max_{\lambda\in\Lambda}\sup_k|\lambda_k| \leqslant 1,$$

and, there exists a constant $C > 0$ such that,

$$\textbf{(A2)} \qquad \sum_{k=1}^{\infty}\sigma_k^4\lambda_k^2 \leqslant C\sum_{k=1}^{\infty}\sigma_k^4\lambda_k^4.$$

Assumptions (A1) and (A2) are rather mild, and they are satisfied in most of the interesting examples. For example, they are trivialy true for projection estimators. Since $|\lambda_k| \leqslant 1$, we also have

$$\sum_{k=1}^{\infty} \sigma_k^4 \lambda_k^4 \leqslant \sum_{k=1}^{\infty} \sigma_k^4 \lambda_k^2,$$

and Assumption (A2) means that both sums are of the same order. The sums $\varepsilon^4 \sum_{k=1}^{\infty} \sigma_k^4 \lambda_k^4$ and $\varepsilon^4 \sum_{k=1}^{\infty} \sigma_k^4 \lambda_k^2$ are the main terms of the variance of $\mathscr{U}(X,\lambda)$.

The Assumption (A1) is quite natural. The first part of (A1) is just to claim that any estimator in Λ has a finite variance. The second point follows from (1.34) the remark that the estimator $\hat{\theta}(\lambda)$ with at least one $\lambda_k \notin [0,1]$ is inadmissible. However, we included the case of negative bounded λ_k since it corresponds to a number of well-known estimators, such as some kernel ones.

Denote

$$\rho(\lambda) = \sup_k \sigma_k^2 |\lambda_k| \left\{ \sum_{k=1}^{\infty} \sigma_k^4 \lambda_k^4 \right\}^{-1/2}$$

and

$$\rho = \max_{\lambda \in \Lambda} \rho(\lambda).$$

Although the main results of this section hold for general ρ, usually think of ρ as being small (for small ε).

Denote also

$$S = \left(\frac{\max_{\lambda \in \Lambda} \sum_{k=1}^{\infty} \sigma_k^4 \lambda_k^2}{\min_{\lambda \in \Lambda} \sum_{k=1}^{\infty} \sigma_k^4 \lambda_k^2} \right)^{1/2},$$

$$M = \sum_{\lambda \in \Lambda} \exp\{-1/\rho(\lambda)\},$$

and

$$L_\Lambda = \log(DS) + \rho^2 \log^2(MS).$$

Note that L_Λ is a term that measure the complexity of the family Λ and not only its cardinality D.

We obtain the following oracle inequality.

Theorem 1.7. *Suppose $\sigma_k \asymp k^\beta$, $\beta \geqslant 0$. Assume that Λ is finite with cardinality D and checking Assumptions (A1)-(A2). There exist constants $\gamma_1, \gamma_2 > 0$ such that for every $\theta \in \ell^2$ and for the estimator $\theta^\star_{ure}$ defined in (1.62), we have for B large enough,*

$$\mathbf{E}_\theta \|\theta^\star_{ure} - \theta\|^2 \leqslant (1 + \gamma_1 B^{-1}) \min_{\lambda \in \Lambda} R(\theta, \lambda) + \gamma_2 B \varepsilon^2 L_\Lambda \; \omega(B^2 L_\Lambda), \qquad (1.63)$$

where

$$\omega(x) = \max_{\lambda \in \Lambda} \sup_k \left(\sigma_k^2 \lambda_k^2 I \left\{ \sum_{i=1}^{\infty} \sigma_i^2 \lambda_i^2 \leqslant x \sup_k \sigma_k^2 \lambda_k^2 \right\} \right), \quad x > 0.$$

Proof. The proof of this theorem may be found in [25].

This result has been extended to the non-compact case in [22].

Function $\omega(x)$ may appear a bit unclear. It depends on the degree of ill-posedness β of the inverse problem and the family of estimators. However, in many examples, it is bounded (up to a constant) by $x^{2\beta}$ (see Examples in [25]). Thus the remainder term in the oracle inequality is usually of order $\varepsilon^2 L_\Lambda^{2\beta+1}$.

By assuming hypothesis on the behaviour of D and S when ε is large, one may obtain an asymptotic exact oracle inequality.

Consider the following family of projection estimators.

Example 1.4. Projection estimators. Let $1 \leqslant N_1 < \ldots < N_D$ be integers. Consider the projection filters $\lambda^s = (\lambda_1^s, \lambda_2^s, \ldots)$ defined by

$$\lambda_k^1 = I(k \leqslant N_1), \quad \lambda_k^2 = I(k \leqslant N_2), \quad \ldots, \quad \lambda_k^D = I(k \leqslant N_D), \quad k = 1, 2, \ldots \quad (1.64)$$

Suppose also, a polynomial behaviour for $S = O(\varepsilon^{-t})$, for some $t > 0$ and $D = O(\varepsilon^{-v})$, for some $v > 0$. We have $\log(DS) = O(\log(1/\varepsilon))$. As noted Assumptions (A1) and (A2) are always true for projection estimators. Note also that here $\omega(x) \leqslant Cx^{2\beta}$ and

$$L_\Lambda \leqslant C \left(\log(DN_D/N_1) + N_1^{-1} \log^2(N_D/N_1) \right).$$

We have the following corollary.

Corollary 1.1. *Assume that* $\Lambda = (\lambda^1, \ldots, \lambda^D)$ *is the set of projection weights defined in (1.64). If* $D = D(\varepsilon)$ *and* $N_1 = N_1(\varepsilon)$, $N_D = N_D(\varepsilon)$ *are such that*

$$\lim_{\varepsilon \to 0} \frac{\log(DN_D/N_1)}{N_1} = 0 \quad (1.65)$$

then for every $\theta \in \ell^2$ *and for the estimator* $\theta_{ure}^\star$ *defined in (1.62), we have*

$$\mathbf{E}_\theta \| \theta_{ure}^\star - \theta \|^2 \leqslant (1 + o(1)) \inf_{\lambda \in \Lambda} R(\theta, \lambda),$$

where $o(1) \to 0$ *uniformly in* $\theta \in \ell^2$.

Proof. The proof of this theorem may be found in [25].

In other words, Corollary 1.1 states that the data-driven selection method $\lambda_{ure}^\star$ behaves itself asymptotically at least as good as the best projection estimator in Λ.

As noted, a major contribution of oracle inequalities is that they usually allow to construct rather easily minimax adaptive estimators. One just has to construct carefully a family of projection estimators which allows to attain the optimal rate of convergence.

Theorem 1.8. *Suppose* $\sigma_k \asymp k^\beta$, $\beta \geqslant 0$. *Assume that* $\Lambda = (\lambda^1, \ldots, \lambda^D)$ *is the set of projection weights defined in (1.64). Choose* $N_j = j$, $j = 1, \ldots, \varepsilon^{-2}$. *Assume that* θ *belongs to the ellipsoid* $\Theta(\alpha, L)$, *where* $a_k = k^\alpha$, $\alpha > 0$, $L > 0$, *defined in (1.40).*

Then the URE estimator θ^*_{ure} defined in (1.62) verifies, for any $\alpha > 0$ and $L > 0$, as $\varepsilon \to 0$,

$$\sup_{\theta \in \Theta(\alpha,L)} \mathbb{E}_\theta \| \theta^*_{ure} - \theta \|^2 \leqslant C \varepsilon^{4\alpha/(2\alpha+2\beta+1)}.$$

This rate is optimal (see Theorem 1.5).

Thus, the URE estimator is then minimax adaptive on the class of ellipsoid.

Proof. The first part of the proof is based on Theorem 1.7. As noted Assumptions (A1) and (A2) are always true for projection estimators. Moreover, here $S = O(\varepsilon^{-2\beta-1})$, $D = O(\varepsilon^{-2})$ and $\omega(x) \leqslant Cx^{2\beta}$. The remainder term is then of order $\varepsilon^2 \log^{2\beta+1}(1/\varepsilon)$.

The second part is just checking that the best projection estimator in Λ attains the optimal rate of convergence. This is true by Theorem 1.6 which gives the optimal choice $N \asymp \varepsilon^{-2/(2\alpha+2\beta+1)}$. Remark also that the remainder term is then much smaller than the optimal rate of convergence.

Remark 1.33. Theorem 1.8 may very easily be modified in order a sharp adaptive estimator, i.e. minimax adaptive which also the exact constant. One just has to replace the projection family by the Pinsker family, which is minimax on ellipsoids (see Theorem 1.5).

Remark 1.34. One may note that even if we have obtained a very precise oracle inequality in Theorem 1.7, the URE method is in fact not so satisfying in simulations. In the case where the problem is really ill-posed, the URE method is in fact not stable enough (see Section 1.3.3.3).

This behaviour, may also be understood, by looking at the results and the proof of Theorem 1.7. These remarks lead to the idea of choosing the bandwidth N by a more stable approach (see Section 1.3.3.2).

Comments

Data-driven choices. One of main difficulties in adaptation or oracle results is that we deal with data-driven choices of N. Thus, the risk of the estimator is very difficult to control since it depends on the observations through X_k and also through the data-driven choice of $\lambda^*(X)$. This really changes the structure of the estimator. For example, a linear estimator $\hat{\theta}(\lambda)$ with a data-driven choice of λ^* is no more linear. The same remark is true for the unbiased risk estimator, which is no more unbiased for a data-driven choice λ^*.

More difficult proofs. This remark is clearly one of the main difficulty when dealing with data-driven choices of N. Thus, adaptive estimator or oracle inequality are usually more difficult to obtain than rates of convergence results for a given estimator.

Proof of an oracle. We will see this influence in the proof of Theorem 1.9. Indeed, one has to deal carefully with remainder terms depending on a data-driven choice λ^*.

The very important following lemma is used in the proofs of Theorems 1.7 and 1.9. It may be found in [25]. It allows to control the deviation of the centered stochastic term. This version is not sharp enough to obtain very precise results (see proof of Theorem 1.9). However, it allows to understand the behaviour of the main stochastic term.

This kind of lemma linked to large deviations and exponential inequalities is usually very important in adaptation or oracle inequality results. One needs to study more carefully the behaviour of the stochastic term, and not only control its variance, which is usually enough in rates of convergences results. These inequalities are also linked to the concentration inequalities, see [124].

Let

$$\bar{\eta}_v = (\sqrt{2}\|v\|)^{-1} \sum_{i=1}^{\infty} v_k(\xi_i^2 - 1)$$

where the sums $\|v\|^2$ and $\sum_{k=1}^{\infty} v_i(\xi_i^2 - 1)$ are understood in the sense of mean squared convergence. Define

$$m(v) = \sup |v_i|/\|v\|.$$

Lemma 1.7. *We have, for* $\kappa > 0$

$$\mathbf{P}(\bar{\eta}_v > x) \leqslant \begin{cases} \exp\left(-\frac{x^2}{2(1+\kappa)}\right) & \text{for } 0 \leqslant \sqrt{2}m(v)x \leqslant \kappa, \\ \exp\left(-\frac{x}{2\sqrt{2}(1+\kappa^{-1})m(v)}\right) & \text{for } \sqrt{2}m(v)x > \kappa. \end{cases} \tag{1.66}$$

Proof. Using the Markov inequality and the formula

$$-\log(1-x) = \sum_{k=1}^{\infty} \frac{x^k}{k}$$

one obtains, for any $0 < t < [\sqrt{2}m(v)]^{-1}$, since $\{\xi_i\}$ are i.i.d. standard Gaussian,

$$\mathbf{P}\{\bar{\eta}_v > x\} \leqslant \exp(-tx)\mathbf{E}\exp(t\bar{\eta}_v)$$

$$= \exp(-tx) \prod_{i=1}^{\infty} \exp\left\{-\frac{tv_i}{\sqrt{2}\|v\|} - \frac{1}{2}\log\left(1 - \frac{\sqrt{2}tv_i}{\|v\|}\right)\right\}$$

$$= \exp(-tx) \exp\left\{\sum_{k=2}^{\infty}\sum_{i=1}^{\infty} \frac{1}{2k}\left(\frac{\sqrt{2}tv_i}{\|v\|}\right)^k\right\}$$

$$= \exp(-tx) \exp\left\{\sum_{k=2}^{\infty} \frac{(\sqrt{2}t)^k}{2k} \sum_{i=1}^{\infty}\left(\frac{v_i}{\|v\|}\right)^2\left(\frac{v_i}{\|v\|}\right)^{k-2}\right\}$$

$$\leqslant \exp(-tx) \exp\left\{\frac{1}{m^2(v)} \sum_{k=2}^{\infty} \frac{1}{2k}[\sqrt{2}tm(v)]^k\right\}$$

$$\leqslant \exp(-tx) \exp\left\{-\frac{1}{2m^2(v)}\log[1 - \sqrt{2}tm(v)] - \frac{t}{\sqrt{2}m(v)}\right\}.$$

Minimization of the last expression with respect to t yields

$$\mathbf{P}\{\bar{\eta}_v > x\} \leqslant \exp\left[\varphi_v(x)\right], \qquad \varphi_v(x) = \frac{1}{2m^2(v)} \log[1 + \sqrt{2}xm(v)] - \frac{x}{\sqrt{2}m(v)}.$$

Note that for $u \geqslant 0$ we have

$$\log(1+u) - u = u \int_0^1 \left(-\frac{\tau u}{1+\tau u}\right) d\tau \leqslant -\int_0^1 \frac{\tau u^2}{1+u} d\tau = -\frac{u^2}{2(1+u)}.$$

Thus

$$\varphi_v(x) \leqslant -\frac{x^2}{2(1 + \sqrt{2}xm(v))},$$

and we obtain

$$\mathbf{P}\{\bar{\eta}_v > x\} \leqslant \exp\left\{-\frac{x^2}{2(1 + \sqrt{2}xm(v))}\right\}, \qquad \forall x > 0. \tag{1.67}$$

It is easy to see that

$$-\frac{x^2}{2(1 + \sqrt{2}xm(v))} \leqslant \begin{cases} -x^2/2(1+\kappa), & \sqrt{2}m(v)x \leqslant \kappa, \\ -x/[2\sqrt{2}(1+\kappa^{-1})m(v)], & \sqrt{2}m(v)x > \kappa. \end{cases}$$

Remark 1.35. There exist two different behaviours for η_v.

The first one is a Gaussian behaviour $\bar{\eta}_v \sim \mathcal{N}(0,1)$, when x is small, i.e. for moderate deviations.

If $\bar{\eta}_v$ was really $\mathcal{N}(0,1)$, the exponential term should be with a constant $1/2$ and not $1/2(1+\kappa)$.

The second behaviour, for large x, i.e. for large deviations, is a Chi-square, centered and dilated by influence of v_i (exponential).

1.3.3.2 Risk Hull Method

In order to present the risk hull minimization, which is an improvement of the URE method, we restrict ourselves to the class of projection estimators. In this case, the URE criterion may be written

$$\mathscr{U}(X,N) = \sum_{k=N+1}^{\infty} (X_k^2 - \varepsilon^2 \sigma_k^2) + \varepsilon^2 \sum_{k=1}^{N} \sigma_k^2.$$

This corresponds in fact to the minimization in N of

$$\bar{R}(X,N) = \sum_{k=N+1}^{\infty} (X_k^2 - \varepsilon^2 \sigma_k^2) + \varepsilon^2 \sum_{k=1}^{N} \sigma_k^2 - \sum_{k=1}^{\infty} (X_k^2 - \varepsilon^2 \sigma_k^2).$$

and then

$$\bar{R}(X,N) = -\sum_{k=1}^{N} X_k^2 + 2\varepsilon^2 \sum_{k=1}^{N} \sigma_k^2.$$

There exists a more general approach which is very close to the URE. This method is called **method of penalized empirical risk**, and in the context of our problem it provides us with the following bandwidth choice

$$N = \arg\min_{N \geqslant 1} \bar{R}_{pen}(X,N), \quad \bar{R}_{pen}(X,N) = \left\{ -\sum_{k=1}^{N} X_k^2 + \varepsilon^2 \sum_{k=1}^{N} \sigma_k^2 + \text{pen}(N) \right\}, \quad (1.68)$$

where pen(N) is a penalty function. The modern literature on this method is very vast and we refer interested reader to [8]. The main idea at the heart of this approach is that severe penalties permit to improve substantially the performance of URE. However, it should be mentioned that the principal difficulty of this method is related to the choice of the penalty function pen(N). In this context, the URE criterion corresponds to a specific penalty called the URE penalty

$$\text{pen}_{ure}(N) = \varepsilon^2 \sum_{k=1}^{N} \sigma_k^2.$$

The idea is usually to choose a heavier penalty, but the choice of such a penalty is a very sensitive problem, and as we will see later, especially in the inverse problems context.

In [26], a more general approach is proposed, called **Risk Hull Minimization** (RHM) which gives a relatively good strategy for the choice of the penalty. The goal is to present heuristic and mathematical justifications of this method.

The heuristic motivation of the RHM approach is based on the oracle approach.

Consider here only the family of projection estimators $\hat{\theta}(N), N \geqslant 1$. Suppose there is an oracle which provides us with θ_k. In this case the oracle bandwidth is evidently given by

$$N_{or} = \arg\min_{N} r(X,N), \quad \text{where } r(X,N) = \|\hat{\theta}(N) - \theta\|^2.$$

This oracle mimimizes the loss and is even better than the oracle of the risk. Let us try to mimic this bandwidth choice. At the first glance this problem seems hopeless since in the decomposition

$$r(X,N) = \sum_{k=N+1}^{\infty} \theta_k^2 + \varepsilon^2 \sum_{k=1}^{N} \sigma_k^2 \xi_k^2, \quad (1.69)$$

neither θ_k^2 nor ξ_k^2 are really known. However, suppose for a moment, that we know all θ_k^2, and try to minimize $r(X,N)$. Since ξ_k^2 are assumed to be unknown, we want to find an upper bound. It means that we minimize the following non-random functional

$$l(\theta, N) = \sum_{k=N+1}^{\infty} \theta_k^2 + V(N), \tag{1.70}$$

where $V(N)$ bounds from above the stochastic term $\varepsilon^2 \sum_{k=1}^{N} \sigma_k^2 \xi_k^2$. It seems natural to choose this function such that

$$\mathbf{E} \sup_N \left[\varepsilon^2 \sum_{k=1}^{N} \sigma_k^2 \xi_k^2 - V(N) \right] \leqslant 0, \tag{1.71}$$

since then we can easily control the risk of any projection estimator with any data-driven bandwidth $N^\star$

$$\mathbf{E}_\theta \| \hat{\theta}(N^\star) - \theta \|^2 \leqslant \mathbf{E}_\theta l(\theta, N^\star). \tag{1.72}$$

This motivation leads to the following definition:

Definition 1.13. A non random function $\ell(\theta, N)$ is called **risk hull** if

$$\mathbf{E}_\theta \sup_N [r(X, N) - \ell(\theta, N)] \leqslant 0.$$

Thus, we can say that $l(\theta, N)$ defined by (1.70) and (1.71) is a risk hull. Evidently, we want to have the upper bound (1.72) as small as possible. So, we are looking for a rather small hull. Note that this hull strongly depends on σ_k^2.

Once $V(N)$ satisfying (1.71) has been chosen, the minimization of $l(\theta, N)$ can be completed by the standard way using the unbiased estimation. Note that our problem is reduced to minimization of $-\sum_{k=1}^{N} \theta_k^2 + V(N)$. Replacing the unknown θ_k^2 by their unbiased estimates $X_k^2 - \varepsilon^2 \sigma_k^2$, we arrive at the following method of adaptive bandwidth choice

$$\bar{N} = \arg\min_N \left[-\sum_{k=1}^{N} X_k^2 + \varepsilon^2 \sum_{k=1}^{N} \sigma_k^2 + V(N) \right].$$

In the framework of the empirical risk minimization in inverse problems, the RHM can be defined as follows. Let the penalty in (1.68) be for any $\alpha > 0$

$$V(N) = \mathrm{pen}_{rhm}(N) = \varepsilon^2 \sum_{k=1}^{N} \sigma_k^2 + (1+\alpha) U_0(N), \tag{1.73}$$

where

$$U_0(N) = \inf \left\{ t > 0 : \ \mathbf{E}(\eta_N I(\eta_N \geqslant t)) \leqslant \varepsilon^2 \sigma_1^2 \right\}, \tag{1.74}$$

with

$$\eta_N = \varepsilon^2 \sum_{k=1}^{N} \sigma_k^2 (\xi_k^2 - 1). \tag{1.75}$$

This RHM penalty corresponds in fact to the URE penalty plus some term $(1 + \alpha) U_0(N)$. One may prove that (see [26]) when $N \to \infty$

$$U_0(N) \approx \left(2\varepsilon^4 \sum_{k=1}^{N} \sigma_k^4 \log \left(\frac{\sum_{k=1}^{N} \sigma_k^4}{2\pi\sigma_1^4} \right) \right)^{1/2}. \tag{1.76}$$

The RHM chooses the bandwidth N_{rhm} according to (1.68) with the penalty function defined by (1.73) and (1.74). The estimator $\theta^\star_{rhm}$ is then defined by

$$\theta_k^\star = I(k \leqslant N_{rhm}) X_k. \tag{1.77}$$

The following oracle inequality provides an upper bound for the mean square risk of this approach.

Theorem 1.9. *Suppose that $\sigma_k \asymp k^\beta$. Let RHM bandwidth choice N_{rhm} according to (1.68) with the penalty function defined by (1.73) and $\theta^\star_{rhm}$ the associated projection estimator defined in (1.77).*
There exist constants $C_* > 0$ and $\delta_0 > 0$ such that for all $\delta \in (0, \delta_0]$ and $\alpha > 1$

$$\mathbf{E}_\theta \| \theta^\star_{rhm} - \theta \|^2 \leqslant (1+\delta) \inf_{N \geqslant 1} R_\alpha(\theta, N) + C_* \varepsilon^2 \left(\frac{1}{\delta^{4\beta+1}} + \frac{1}{\alpha - 1} \right), \tag{1.78}$$

where

$$R_\alpha(\theta, N) = \sum_{k=N+1}^{\infty} \theta_k^2 + \varepsilon^2 \sum_{k=1}^{N} \sigma_k^2 + (1+\alpha) U_0(N).$$

Proof. Many of the details are deleted, in order to keep only the idea behind the risk hull. The proof in its full length can be found in [26].

The proof is now in two parts:
The first part is to prove the following lemma.

Lemma 1.8. *We have, for any $\alpha > 0$,*

$$l_\alpha(\theta, N) = \sum_{k=N+1}^{\infty} \theta_k^2 + \varepsilon^2 \sum_{k=1}^{N} \sigma_k^2 + (1+\alpha) U_0(N) + \frac{C\varepsilon^2}{\alpha}.$$

is a risk hull, where $C > 0$ is a positive constant.

Proof. Using (1.69) and (1.70), remark that

$$\mathbf{E} \sup_N (\eta_N - (1+\alpha) U_0(N))_+ \leqslant \frac{C\varepsilon^2}{\alpha}$$

implies

$$\mathbf{E}_\theta \sup_N (r(X, N) - l_\alpha(\theta, N))_+ \leqslant 0.$$

We have

$$\mathbf{E} \sup_N (\eta_N - (1+\alpha) U_0(N))_+ \leqslant \sum_{N=1}^{\infty} \mathbf{E} (\eta_N - (1+\alpha) U_0(N))_+. \tag{1.79}$$

The definition of $U_0(N)$ in (1.74) implies

$$\mathbf{E}\left(\eta_N - U_0(N)\right)_+ \leqslant \mathbf{E}\left(\eta_N I(\eta_N \geqslant U_0(N))\right) \leqslant \varepsilon^2 \sigma_1^2.$$

Moreover, by integrating by parts we obtain

$$\mathbf{E}\left(\eta_N - (1+\alpha)U_0(N)\right)_+ = \int_{(1+\alpha)U_0(N)}^{\infty} \mathbf{P}(\eta_N > x)dx. \tag{1.80}$$

Denote by

$$M_N = \varepsilon^2 \max_{k=1,\dots,N} \sigma_k^2$$

and

$$\Sigma_N = \varepsilon^4 \sum_{k=1}^{N} \sigma_k^4.$$

Since the inverse problem is mildly ill-posed, one obtains, by use of (1.46), as $N \to \infty$,

$$M_N \asymp \varepsilon^2 N^{2\beta}, \tag{1.81}$$

$$\Sigma_N \asymp \varepsilon^2 \sum_{k=1}^{N} k^{4\beta} \asymp \varepsilon^4 N^{4\beta+1} \tag{1.82}$$

and, using (1.76),

$$U_0(N) \asymp \varepsilon^2 N^{2\beta+1/2} \sqrt{\log N}. \tag{1.83}$$

Considering only the family of projection estimators, we get another version of Lemma 1.7 with $\kappa = 1/4$.

Lemma 1.9. *We have*

$$\mathbf{P}(\eta_N > x) \leqslant \begin{cases} \exp\left(-\frac{x^2}{5\Sigma_N}\right) & 0 \leqslant x \leqslant \frac{\Sigma_N}{4M_N}, \\ \exp\left(-\frac{x}{20M_N}\right) & x > \frac{\Sigma_N}{4M_N}. \end{cases} \tag{1.84}$$

Remark that, due to (1.81)-(1.83), $U_0(N) \leqslant \Sigma_N/4M_N$, when N is large.
We can then divide in two parts the integral in (1.80),

$$\int_{(1+\alpha)U_0(N)}^{\infty} \mathbf{P}(\eta_N > x)dx =$$

$$= \int_{(1+\alpha)U_0(N)}^{\frac{\Sigma_N}{4M_N}} \mathbf{P}(\eta_N > x)dx + \int_{\frac{\Sigma_N}{4M_N}}^{\infty} \mathbf{P}(\eta_N > x)dx. \tag{1.85}$$

When $x > \Sigma_N/4M_N$, we have, when $N \to \infty$,

$$\int_{\Sigma_N/4M_N}^{\infty} \exp\left(-\frac{x}{20M_N}\right)dx \leqslant CM_N \exp\left(-C\frac{\Sigma_N}{M_N^2}\right) \asymp C\varepsilon^2 N^{2\beta} \exp\left(-CN\right). \tag{1.86}$$

Moreover

$$\int_{(1+\alpha)U_0(N)}^{\infty} \exp\left(-\frac{x^2}{5\Sigma_N}\right) dx \leq \int_{(1+\alpha)U_0(N)}^{\infty} \frac{x}{(1+\alpha)U_0(N)} \exp\left(-\frac{x^2}{5\Sigma_N}\right) dx$$

$$\leq \frac{5\Sigma_N}{2(1+\alpha)U_0(N)} \exp\left(-\frac{(1+\alpha)^2 U_0(N)^2}{5\Sigma_N}\right).$$

Thus, using (1.76), we obtain

$$\int_{(1+\alpha)U_0(N)}^{\infty} \exp\left(-\frac{x^2}{5\Sigma_N}\right) dx$$

$$\leq C\sqrt{\Sigma_N} \exp\left(-\frac{2}{5}(1+\alpha)^2 \log\left(\frac{\sum_{k=1}^{N} \sigma_k^4}{2\pi\sigma_1^4}\right)\right). \tag{1.87}$$

Using (1.82), remark that the term in (1.86) is smaller than the one in (1.87), as $N \to \infty$. Using (1.79), (1.80) and (1.85), we then obtain

$$\mathbf{E} \sup_{N} (\eta_N - (1+\alpha)U_0(N))_+ \leq \sum_{N=1}^{\infty} C\varepsilon^2 \exp\left(-\left(\frac{2}{5}(1+\alpha)^2 - \frac{1}{2}\right)\log(N)\right).$$

Thus, for α large enough ($\alpha > 2$), the term is then summable in N and we obtain

$$\mathbf{E} \sup_{N} (\eta_N - (1+\alpha)U_0(N))_+ \leq \frac{C\varepsilon^2}{\alpha}.$$

The proof for $\alpha > 0$ small is much more technical and based on chaining arguments (see [26]).

In the second part of the proof of Theorem 1.9, we need to prove that we are able to minimize this risk hull based on the data. Since $l_\mu(\theta, N)$ is a risk hull for any $\mu > 0$ we have

$$l_\mu(\theta, N) = \sum_{k=N+1}^{\infty} \theta_k^2 + \varepsilon^2 \sum_{k=1}^{N} \sigma_k^2 + (1+\mu)U_0(N) + \frac{C\varepsilon^2}{\mu}, \tag{1.88}$$

and therefore

$$\mathbf{E}_\theta \|\hat{\theta}(N_{rhm}) - \theta\|^2 \leq \mathbf{E}_\theta l_\mu(\theta, N_{rhm}). \tag{1.89}$$

On the other hand, since N_{rhm} minimizes $\bar{R}_{pen}(X, N)$, we have for any integer N

$$\mathbf{E}_\theta \bar{R}_{pen}(X, N_{rhm}) \leq \mathbf{E}_\theta \bar{R}_{pen}(X, N) = R_\alpha(\theta, N) - \|\theta\|^2. \tag{1.90}$$

In order to combine the inequalities (1.89) and (1.90), we rewrite $l_\mu(\theta, N_{rhm})$ in terms of $\bar{R}_{pen}(X, N_{rhm})$,

$$l_\mu(\theta, N_{rhm}) = \bar{R}_{pen}(X, N_{rhm}) + \|\theta\|^2 + \frac{C\varepsilon^2}{\mu}$$

$$+2\varepsilon \sum_{k=1}^{N_{rhm}} \sigma_k \theta_k \xi_k + \varepsilon^2 \sum_{k=1}^{N_{rhm}} \sigma_k^2 (\xi_k^2 - 1) - (\alpha - \mu) U_0(N_{rhm}).$$

Therefore, using this equation, (1.89) and (1.90), we obtain that for any integer N

$$\mathbf{E}_\theta \|\hat{\theta}(N_{rhm}) - \theta\|^2 \leqslant R_\alpha(\theta, N) + \frac{C\varepsilon^2}{\mu} + \mathbf{E}_\theta 2\varepsilon \sum_{k=1}^{N_{rhm}} \sigma_k \theta_k \xi_k$$

$$+ \mathbf{E}_\theta \left[\varepsilon^2 \sum_{k=1}^{N_{rhm}} \sigma_k^2 (\xi_k^2 - 1) - (\alpha - \mu) U_0(N_{rhm}) \right].$$

The next step is to control the last two terms in the above equation. This part of proof is not done here (see [26]).

This control should be done for any data-driven choice N^* (or N_{rhm}), this is why these terms are difficult to control. Moreover, to get a sharp oracle inequality, one has to be rather precise.

The first term, of the last two terms, may be included in the left term (the risk of the RHM estimator) and in the remainder term. However, this part of the proof is one of the more delicate. One really has to control this stochastic term for any data-driven N^* (see [26]).

The second term, of the last two terms, is very close to Lemma 1.8 and its proof. Thus, we may use again the risk hull in order to control it.

As noted, a major contribution of oracle inequalities is that they usually allow to construct rather easily minimax adaptive estimators. Here the proof is very simple because the family of estimators corresponds to all possible choices of N.

Theorem 1.10. *Suppose $\sigma_k \asymp k^\beta$, $\beta \geqslant 0$. Let RHM bandwidth choice N_{rhm} according to (1.68) with the penalty function defined by (1.73) and θ_{rhm}^* the associated projection estimator defined in (1.77).*

Assume that θ belongs to the ellipsoid $\Theta(\alpha, L)$, where $a_k \asymp k^\alpha$, $\alpha > 0$, $L > 0$, defined in (1.40). Then the RHM estimator θ_{rhm}^* verifies, for any $\alpha > 0$ and $L > 0$, as $\varepsilon \to 0$,

$$\sup_{\theta \in \Theta(\alpha, L)} \mathbf{E}_\theta \|\theta_{rhm}^* - \theta\|^2 \leqslant C \varepsilon^{4\alpha/(2\alpha + 2\beta + 1)}.$$

This rate is optimal (see Theorem 1.5).

Thus, the RHM estimator is then minimax adaptive on the class of ellipsoid.

Proof. The proof is a direct consequence of Theorem 1.6 and 1.9. One has to note that, due to (1.83), $U_0(N) = o(\varepsilon^2 \sum_{k=1}^N \sigma_k^2)$ as $N \to \infty$. Asymptotically, the RHM penalty is negligible as compared to the URE penalty. Thus, the penalized oracle $R_\alpha(\theta, N)$ on the right hand side of Theorem 1.9 still attains the optimal rate of convergence.

Remark 1.36. In order to construct a sharp adaptive estimator on ellipsoids, one has to obtain results for the Pinsker family. The RHM method has been extended to the Pinsker family in [99].

Comments

Penalized oracle. We have an oracle inequality but with a penalty term on the RHS. This is usually called a (penalized) oracle inequality. This is standard in the penalized empirical risk approach. At the first sight, the result may look weaker than in Theorem 1.7. Indeed, the main term is a penalized oracle here when it was the true oracle in Theorem 1.7. However, here the remainder term is better. In Theorem 1.7, the remainder term depends on the cardinality and on the complexity of the family of estimators. In Theorem 1.9, there is no such price, and moreover the family may be infinite. However, as will be explained by simulations in Section 1.3.3.3, the even more important point is that the constant is much more under control than in Theorem 1.7.

Natural penalty. By (1.76) the penalty $U_0(N)$ is almost of the order of the standard deviation of the empirical risk. This seems rather natural, since it really controls the behaviour of the empirical risk, i.e. not only its expectation but also its standard deviation.

Second order penalty. We have $U_0(N) = o(\varepsilon^2 \sum_{k=1}^N \sigma_k^2)$ as $N \to \infty$, since (1.83). We add a penalty (see (1.73)) which is small compared to the URE penalty. In fact, the RHM penalty may be thought as the URE penalty plus a second order penalty. From an asymptotical point of view, there is no real difference between the URE and the RHM. Thus, the two methods should be very close. A consequence of the previous remark, is that the (penalized) oracle inequality is then (asymptotically) as sharp as the one in Theorem 1.7. Asymptotically, one may obtain exactly the same results, since the penalty is smaller. Thus, minimax adaptive estimators may be constructed directly (see Theorem 1.10).

Direct problem. In the direct problem $(A = I)$, i.e. in Gaussian white noise model, due to (1.76), the penalty is then:

$$Pen_{rhm}(N) = \varepsilon^2 N + (1 + \alpha)U_0(N),$$

where

$$U_0(N) \approx \left(2\varepsilon^4 N \log \frac{N}{2\pi} \right)^{1/2}.$$

One may see that we really add a second order penalty.

Difference between RHM and URE. On the one hand, the previous remarks show that the RHM penalty is equal to the URE penalty plus a small term (compared to the URE penalty). On the other hand, there exist main differences between the two estimators, especially in the case of inverse problems. RHM is much more stable than URE (see Section 1.3.3.3). Moreover, in the simulations, it is always more accurate, even in the direct problem. However, the difference is much more important in ill-posed framework.

Asymptotics in inverse problems. One of the reason for its instability is that URE is based on some asymptotical ideas. In inverse problems, usually N is not very large, due to the increasing noise. Indeed, in the ill-posed context, the term $\sigma_k \to \infty$. It means that the noise is really increasing with k. One has to be very careful with high frequencies. More or less, it is very difficult to choose a large number of coefficients N. On the one hand, the minimax theory, claims that the optimal choice of N is going to infinity in nonparametric statistics (see for example Theorem 1.6). On the other hand, the choice of N cannot be too large, otherwise, in real inverse problems the noise will explose.

Thus, one has to be very careful with asymptotics in inverse problems.

Penalty computed by Monte Carlo. The penalty $U_0(N)$ may be computed by Monte Carlo simulations. Indeed, the definition of $U_0(N)$ in (1.74) has no explicit solution. There exists an approximation of $U_0(N)$ in (1.76), but it is true for N large enough. As noted, N is not so large in inverse problems. Thus, a more careful and accurate way to compute $U_0(N)$ is by use of Monte Carlo. It is a bit time consuming, but it is done only once for one given inverse problem.

Explicit penalty. By use of RHM we obtain, an explicit penalty which comes from the proof of Theorem 1.9. It is really by looking inside the proof of Lemma 1.8 that one may understand the penalty form. The constraint, that one wants to have a risk hull really help in choosing such a penalty.

Another very important point, is that after, this penalty may be used directly in simulations. The method, really gives, an explicit penalty. There is no gap between the penalty needed in Theorem 1.9 and the one used in the simulation study.

1.3.3.3 Simulations

In order to illustrate the difference between direct and inverse estimation, we will carry out a very simple numerical experiment. Obviously, we cannot compute it for all $\theta \in \ell^2$. Therefore, let us take $\theta_k \equiv 0$ and compute the ratio between the risk and the risk of the oracle for two cases $\sigma_k \equiv 1$ and $\sigma_k = k$. The first case corresponds to classical function estimation (direct estimation), whereas the second is related to the estimation of the first order derivative of a function (inverse estimation). Notice that in both cases the risk of the oracle is clearly $\inf_N R(0,N) = \varepsilon^2$ since $\mathrm{argmin}_N R(0,N) = 1$.

In order to study the performance of the URE, we generate 2000 independent random vectors of y^j, $j = 1, ..., 2000$ with the components defined by (1.5). For each vector we compute $N_{ure}(y^j)$ and the normalized error $\|\hat{\theta}[N_{ure}(y^j)] - \theta\|^2/\varepsilon^2$ and plot these values as a stem diagram. We also compute the mean empirical bandwidth N_{emp} and the normalized mean empirical risk R_{emp} by

$$N_{emp} = \frac{1}{2000} \sum_{j=1}^{2000} N_{ure}(y^j), \quad R_{emp} = \frac{1}{2000\varepsilon^2} \sum_{j=1}^{2000} \|\hat{\theta}[N_{ure}(y^j)] - \theta\|^2.$$

Let us discuss briefly the numerical results of this experiment shown on Figure 1.1. The first display (direct estimation) shows that the URE method works reasonably well. Almost all bandwidths $N_{ure}(y^j)$ are relatively small (their mean is 1.98)

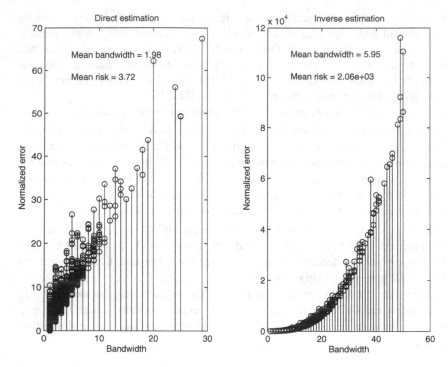

Fig. 1.1 The method of unbiased risk estimation

and the normalized error is 3.72. The second display shows that the distribution of $N_{ure}(y^j)$ changed essentially. Now the mean bandwidth is 5.95 and there are sufficiently many bandwidths $N_{ure}(y^j)$ greater than 20. This results in a catastrophic normalized error around 2000.

In this section, we present some numerical properties of the RHM approach. We will study in a more general context than the previous no-signal one, i.e. $\theta_k \equiv 0$. Numerical testing of nonparametric statistical methods is a very difficult and delicate problem. The goal of this section is to illustrate graphically Theorem 1.7 and Theorem 1.9. To do that, we propose to measure statistical performance of a method $N^\star$ by its *oracle efficiency* defined by

$$
e_{or}(\theta, N^\star) = \frac{\inf_N \mathbf{E}_\theta \|\hat{\theta}(N) - \theta\|^2}{\mathbf{E}_\theta \|\hat{\theta}(N^\star) - \theta\|^2}.
$$

If the oracle efficiency of a method is close to 1 then the risk is very close to the risk of the oracle.

It should be mentioned that we use the inverse of the previous ratio since we want to get a good graphical representation of the performance. We have just seen in the previous part that the ratio can vary from 1 to 2000 for the URE method. This results in a degenerate plot. Therefore, in order to avoid this effect, we use this definition of the oracle efficiency $e_{or}(\theta, N^*)$.

Since it is evidently impossible to compute the oracle efficiency for all $\theta \in \ell^2$, we choose a sufficiently representative family of vectors θ. In what follows we will use the following family, with polynomial decreasing,

$$\theta_k^a = \frac{a\varepsilon}{1 + (k/W)^m},$$

where ε is the noise level, a is called amplitude, W bandwidth, and m smoothness.

We shall vary a in a large range and plot $e_{or}(\theta^a, N^*)$ as a function of a which is directly related with the signal-to-noise ratio in the model considered. In a statistical framework a^2 would be n the number of observations. The parameters $m = 6$ and $W = 6$ are fixed. Many other examples of (W, m) were looked at, simulations showed that the oracle efficiency exhibits similar behaviour.

Two methods of data-driven bandwidth choice will be compared: the URE and the RHM with $\alpha = 1.1$. One may note that for these methods $e_{or}(\theta^a, N^*)$ does not depend on ε. This function was computed by the Monte Carlo method with 40000 replications.

We start with the direct estimation where $\sigma_k \equiv 1$. Figure 1.2 shows the oracle efficiency of the URE (left panel) and the oracle efficiency of the RHM (right panel). Comparing these plots, one can say that both methods work reasonably well. Both efficiencies are very close to 1. The risk of URE method is around $1/0.75 = 1.33$ times the risk of oracle, when RHM method is around $1/0.82 = 1.22$ times the oracle. Thus RHM is always better than URE, but the ratio is something like 5% to 10%.

However, if we deal with an inverse problem ($\sigma_k = k$), we can already see a significant difference between these methods. The corresponding oracle efficiencies are plotted on the left and on the right panels of Figure 1.3. For small values a the performance of the URE is very poor, whereas the RHM demonstrates a very stable behaviour. For very large $a = 500$ the oracle efficiency of the URE is of order 0.16, which means that its risk is around 6 times the one of the oracle. For smaller $a = 100$, it is around 10 times the oracle. In the meantime, the RHM has always an efficiency greater than 0.4 and usually around 0.5, i.e. 2 times the risk of the oracle.

The last Figure 1.4 deals with the case when the inverse problem is more ill-posed ($\sigma_k = k^2$). In this situation the URE fails completely. Its maximal oracle efficiency is of order $3*10^{-4}$, i.e. 10000 times the oracle. Nevertheless, the RHM has a good efficiency (greater than 0.3). Its risk is then around 3 times those of the oracle.

Another remark is that the RHM is really stable compared to the increasing degree of ill-posedness β of the problem. The efficiency is worse when the inverse problem is more difficult, but it is always reasonable. The behaviour of URE is completely different, it really explodes with β.

One may also see that URE is really based on asymptotic ideas. Indeed, its oracle efficiency is highly increasing with the amplitude a. On the other hand, RHM is stable, and does not rely on large values of a.

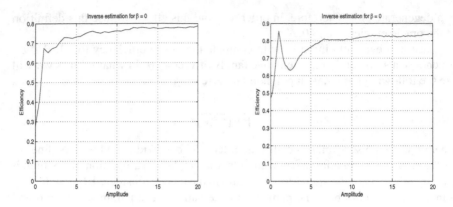

Fig. 1.2 Oracle efficiency of URE (left) and RHM (right) for direct estimation.

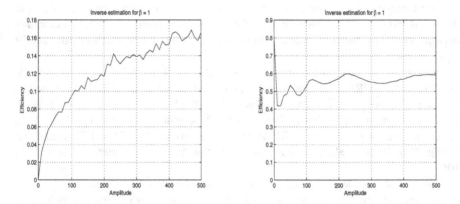

Fig. 1.3 Oracle efficiency of URE and of RHM for inverse problem ($\beta = 1$).

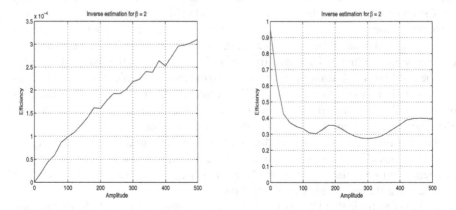

Fig. 1.4 Oracle efficiency of URE and of RHM for inverse problem ($\beta = 2$).

This simulation study, shows that there is a huge difference between the two methods, at least in inverse problems. This may be surprising, since the RHM penalty, was supposed to be of second order. Then, the two methods should be closely related. However, this point of view, mainly relies on asymptotic ideas. As noted before, in inverse problems, one has to be really careful with asymptotics. This may really be seen here, where these two methods have a very different behaviour.

In the context of Theorem 1.7 and Theorem 1.9, this example shows also that the constants which appear in the remainder terms are quite different. The one in Theorem 1.9, seems to be small and really under control. While the one in Theorem 1.7, C_* is in fact really large. Unfortunately, it means that the terms which are asymptotically small in Theorem 1.7 may dominate the risk of oracle.

1.3.4 Universal Optimality

1.3.4.1 Blockwise Estimators

In this section, we present a more general approach to optimality. Namely, we construct a sequence of weights λ_{pbs}^* such that the penalized blockwise Stein estimator $\theta_{pbs}^* = \hat{\theta}(\lambda_{pbs}^*)$ satisfies both some exact oracle inequalities (for typical examples of classes Λ) and the (sharp) minimax adaptivity property (Definition 1.12) (for a large scale of classes Θ_α).

An important fact is that the estimator does not belong to either of the typical classes Λ but it outperforms the oracles λ_0 corresponding to these classes. This property can be called **universal optimality** over a large scale of classes Λ. This point of view is different from the model selection ideas in Section 1.3.3, where the data-driven choice take its values in the family Λ.

An important point here, is to find a large family Λ in order to obtain oracle inequalities valid for many different estimators. The first step is close to the approach of unbiased risk estimation. Indeed, one would like to minimize the criteria $\mathscr{U}(X,\lambda)$ on such a family.

What is the reasonable set of λ where the minimization of $\mathscr{U}(X,\lambda)$ should be done? Minimizing $\mathscr{U}(X,\lambda)$ with respect to all possible λ yields $\lambda_k = (1 - \varepsilon^2/X_k^2)$ or $\lambda_k = (1 - \varepsilon^2/X_k^2)_+$ if we restrict the minimization to $\lambda_k \in [0,1]$. It is easy to see that the risk of the estimator $\{\lambda_k X_k\}$ is diverging if the sum is taken over all k and is at least as great as $\varepsilon^2 N$ if one considers the sum over $k \leqslant N$ for some integer N in the definition of $\mathscr{U}(X,\lambda)$. Since N should be chosen in advance, such an estimator has poor adaptation properties, and minimizing over all λ makes no sense.

A more fruitful idea is to minimize $\mathscr{U}(X,\lambda)$ in a restricted class of sequences, for example over one of the classes Λ discussed in Section 1.2.2.

Choose λ^* as a minimizer of $\mathscr{U}(X,\lambda)$ over $\lambda \in \Lambda$ in order to mimic the linear oracle on Λ. However, this principle is difficult to apply for huge classes, such as Λ_{mon}, the class of monotone weights. [59] suggests the minimization of $\mathscr{U}(X,\lambda)$ on a truncated version of the class Λ_{mon}.

The search for more economic but yet huge enough subclasses of weight sequences λ leads in particular to the family of blockwise constant weights which can interpreted as sieves over various sets of λ. Blockwise constant weights have been discussed in statistical literature starting from [44], and more recently by [47, 105]; for wavelets, see [40, 78, 67].

The key feature of our estimator is that it "mimics" the *monotone oracle* λ_0^{mon} defined as a solution of

$$R(\theta, \lambda_0^{mon}) = \min_{\lambda \in \Lambda_{mon}} R(\theta, \lambda), \tag{1.91}$$

where Λ_{mon} is the class of monotone sequences. Consider the class of monotone weights sequences

$$\Lambda_{mon} = \{\lambda = \{\lambda_k\} \in \ell^2 : 1 \geqslant \lambda_1 \geqslant \ldots \geqslant \lambda_k \ldots \geqslant 0\},$$

and the class of *monotone estimators*

$$\hat{\theta}_k = \lambda_k X_k,$$

where $\{\lambda_k\} \in \Lambda_{mon}$ and X_k is defined in (1.6).

If the coefficients θ_k are monotone non-increasing, remark that the monotone oracle is equal to the linear oracle.

Restrict the attention to the class Λ_{mon} since it contains the most interesting examples of weight sequences $\{\lambda_k\}$. The projection weights and the Tikhonov weights belong to Λ_{mon} (see Section 1.2.2). Next, typically σ_k are monotone non-decreasing and a_k in the definition of the ellipsoid in (1.40) are monotone non-decreasing. The Pinsker weights also belong to Λ_{mon}. It can be shown that some minimax solutions on other bodies in ℓ^2 than ellipsoids (e.g. parallelepipeds) are also in Λ_{mon}, see [31].

We are looking for an adaptive estimator $\theta^\star = (\theta_1^\star, \theta_2^\star, \ldots)$ of the form

$$\theta_k^\star = \lambda_k^\star X_k,$$

where $\lambda_k^\star$ are some data-driven weights.

A well-known idea of choosing $\lambda^\star$ is based on the unbiased estimation of the risk by minimizing criteria $\mathcal{U}(X, \lambda)$ defined in (1.59) among the family Λ (see Section 1.3.3.1). The difference here is that the class Λ is not some given class of estimators (projection, Tikhonov,...) but the very large class Λ_{mon} of monotone estimators.

However, as noted before, this class Λ_{mon} is maybe too large. Consider instead, the class Λ_b of coefficients with piecewise constant λ_k over suitably chosen blocks.

Define the class of **blockwise estimators**

$$\hat{\theta}_k = \lambda_k X_k,$$

where $\lambda \in \Lambda_b$ is the set of piecewise constant sequences,

$$\Lambda_b = \{\lambda \in \ell^2 : 0 \leqslant \lambda_k \leqslant 1, \lambda_k = \lambda_{\kappa_j}, \forall k \in I_j, \lambda_k = 0, k > N_{max}\},$$

where I_j denote the block $I_j = \{k \in [\kappa_{j-1}, \kappa_j - 1]\}, j = 0, \ldots, J-1$ and J, N_{max}, $\kappa_j, \ j = 0, \ldots, J$, are integers such that $\kappa_0 = 1, \ \kappa_J = N_{max} + 1, \ \kappa_j > \kappa_{j-1}$.

Denote also by $T_j = \kappa_j - \kappa_{j-1}$ the size of the blocks I_j, for $j = 1, \ldots, J$.

1.3.4.2 Stein's Estimator

In this section, we change to a slightly different model in order to present and discuss the so-called Stein phenomenon. This problem goes back to the work of [120], and has been extended since. However, it is still one of the most surprising result in statistics. This section is based on [18, 129].

Consider the following model, which is a finite version of the sequence space model in the direct case (i.e. $b_k \equiv 1$),

$$y_k = \theta_k + \varepsilon \xi_k, \ k = 1, \ldots, d, \tag{1.92}$$

where d is some integer, $\{\xi_k\}$ are i.i.d. $\mathcal{N}(0,1)$. The statistical problem is to estimate θ based on the data y.

In this simple situation, the Maximum Likelihood Estimator is then $\hat{\theta}_{mle} = y$. This estimator was believed, to be the best possible estimator in this context. Its risk is

$$R(\theta, \hat{\theta}_{mle}) = \varepsilon^2 d, \ \forall \theta \in \mathbb{R}^d.$$

However, [120] discovered a very strange phenomenon. Indeed, he constructed an estimator, the **Stein estimator**

$$\hat{\theta}_S = \left(1 - \frac{\varepsilon^2 d}{\|y\|^2}\right) y, \tag{1.93}$$

for which we have, see proof of Lemma 3.10 in [129],

$$\mathbf{E}_\theta \|\hat{\theta}_S - 0\|^2 = \varepsilon^2 d - \varepsilon^4 d(d-4) \mathbf{E}_\theta \left(\frac{1}{\|y\|^2}\right),$$

and

$$\mathbf{E}_\theta \|\hat{\theta}_S - \theta\|^2 \leqslant \varepsilon^2 d - \frac{\varepsilon^4 d(d-4)}{\|\theta\|^2 + \varepsilon^2 d}. \tag{1.94}$$

Thus, the main result in [120] is that if $d \geqslant 5$,

$$\mathbf{E}_\theta \|\hat{\theta}_S - \theta\|^2 < \mathbf{E}_\theta \|y - \theta\|^2, \ \forall \theta \in \mathbb{R}^d.$$

This very surprising result proves that the MLE estimator y is not even admissible (for $d \geqslant 5$).

Written in a slightly different framework, [120] discovered that the Stein estimator is better at each point $\theta \in \mathbb{R}^d$ than the mean $\bar{X}$ (for $d \geqslant 5$).

Looking carefully at (1.94), note that the improvement on y is by a constant at point $\theta_k \equiv 0$, but also if $\|\theta\| \asymp \varepsilon$. However, when θ is larger, the improvement is of second order. Nevertheless, this is an asymptotical point of view, and the gain is valid for any $\theta \in \mathbb{R}^d$.

Several versions of the Stein estimator have been defined since then, many of them which improved on the basic estimator. One example is the **positive Stein estimator**

$$\hat{\theta}_s = \left(1 - \frac{\varepsilon^2 d}{\|y\|^2}\right)_+ y. \tag{1.95}$$

The following result may be found in Lemma 3.9 in [129], for all $d \geqslant 1$,

$$\mathbf{E}_\theta \|\hat{\theta}_s - \theta\|^2 < \mathbf{E}_\theta \|\hat{\theta}_S - \theta\|^2, \ \forall \theta \in \mathbb{R}^d.$$

Another famous version is the **James-Stein estimator** (and its positive version),

$$\hat{\theta}_{JS} = \left(1 - \frac{\varepsilon^2(d-2)}{\|y\|^2}\right) y,$$

see [76], which is better than the MLE estimator y even for $d \geqslant 3$.

Remark 1.37. The (positive) Stein estimator has an effect, even if still very surprising, which may be understood. On the one hand, when the whole signal $\|y\|^2$ is large (compared to $\varepsilon^2 d$), then one may rely on the data, and estimate θ by something very close to y. On the other hand, if $\|y\|^2$ is small, then one estimates θ by something close to 0, or even equal to 0 if $\|y\|^2 \leqslant \varepsilon^2 d$. The information on the whole sequence $\{y_k\}$ helps in estimating a single coefficient θ_k in a better way than just by using y_k.

Moreover, the Stein estimator has a role of moving the data y to 0 by some factor. This effect is known nowadays as the *Stein shrinkage*. The idea is that one shrinks the observations, more or less, towards 0 in order to improve on y.

The ideas of Stein have been very successful and popular among statisticians. Moreover, since Stein's result is valid in large dimensions d, his ideas are still the topic of a vast literature in nonparametric statistics, where d is very large, even infinite, see [41, 14, 77, 17, 31, 18, 97, 114, 129]. The main common point among these papers, is to try to estimate the infinite sequence $\{\theta_k\}$ by using block estimators, as sieves, see Section 1.3.4.1. Then on each of these blocks, the idea is to estimate the coefficients θ_k by use of the Stein estimator. As already noted, the nonparametric context is well suited, since then the blocks will be large, and Stein's estimator successful. One of the main difficulties is then related to the choice of the size of the blocks.

1.3.4.3 Blockwise Stein's Rules

The construction of the estimator θ^*_{pbs} is the following.

Divide the set of coefficients θ_k into blocks in a proper way, and apply a penalized version of Stein's estimator on each block. The penalty should be rather small but non-zero. The same construction with non-penalized Stein's estimators can be also implemented, but leads to more limited results (see [31]).

Note that the solution $\lambda_{bs}^\star$ of the minimization problem

$$\mathcal{U}(X, \lambda_{bs}^\star) = \min_{\lambda \in \Lambda_b} \mathcal{U}(X, \lambda)$$

is given by $\lambda_{bs}^\star = (\lambda_1^\star, \lambda_2^\star, \dots)$, where

$$\lambda_k^\star = \begin{cases} \left(1 - \dfrac{\Gamma_{(j)}^2}{\|X\|_{(j)}^2}\right)_+ , & k \in I_j, \ j = 1, \dots, J, \\ 0 , & k > N_{max}, \end{cases} \tag{1.96}$$

with $x_+ = \max(0, x)$,

$$\Gamma_{(j)}^2 = \varepsilon^2 \sum_{k \in I_j} \sigma_k^2, \qquad \|X\|_{(j)}^2 = \sum_{k \in I_j} X_k^2,$$

and

$$\Delta_{(j)} = \frac{\max_{k \in I_j} \sigma_k^2}{\sum_{k \in I_j} \sigma_k^2}.$$

The weights (1.96) define a *blockwise Stein rule*. The **blockwise Stein estimator** is

$$\theta_k^\star = \lambda_k^\star X_k,$$

where $\lambda_{bs}^\star = \{\lambda_k^\star\}$ is defined in (1.96).

However, for mildly ill-posed inverse problems, the estimator $\theta_{bs}^\star$ can be modified to have better properties.

We now modify the weights $\lambda_{bs}^\star$ and define $\lambda_{pbs}^\star$ by

$$\lambda_k^\star = \begin{cases} \left(1 - \dfrac{\Gamma_{(j)}^2 (1 + p_j)}{\|X\|_{(j)}^2}\right)_+ , & k \in I_j, \ j = 1, \dots, J, \\ 0 , & k > N_{max}, \end{cases}$$

where $0 \leqslant p_j \leqslant 1$ is some penalty term.

Finally, the estimator has the form $\theta_{pbs}^\star = (\theta_1^\star, \theta_2^\star, \dots)$ where

$$\theta_k^\star = \begin{cases} \left(1 - \dfrac{\Gamma_{(j)}^2 (1 + p_j)}{\|X\|_{(j)}^2}\right)_+ X_k , & k \in I_j, \ j = 1, \dots, J, \\ 0 , & k > N_{max}. \end{cases} \tag{1.97}$$

This estimator is called the **penalized blockwise Stein estimator**.

Remark 1.38. The penalizing factor $(1 + p_j)$ forces the estimator to contain fewer nonzero coefficients $\theta_k^\star$ than for the usual blockwise Stein's rule (1.96): our estimator is more "sparse". The general choice of the penalty p_j will be $p_j = \Delta_j^a$, where $0 < a < 1/2$. The assumption $a < 1/2$ is important. Intuitively, this effect is easy to explain. If b_k decreases as a power of k we have:

$$\text{standard deviation}(Z_j)/\text{expectation}(Z_j) \asymp \Delta_{(j)}^{1/2}$$

where Z_j is the stochastic error term corresponding to jth block. Hence, to control the variability of stochastic terms, one needs a penalty that is slightly larger than $\Delta_{(j)}^{1/2}$.

More general penalties are presented in [31].

1.3.4.4 Construction of Blocks

Introduce now a *special construction of blocks* I_j which may be called *weakly geometrically increasing blocks*. In Theorem 1.11 we will show that with this construction the penalized blockwise Stein estimator verifies an oracle inequality. This construction (or some versions of it) is used by [105] but also in [58, 32, 114].

Let ν_ε be an integer valued function of ε such that $\nu_\varepsilon \geqslant 5$ and $\nu_\varepsilon \to \infty$ as $\varepsilon \to 0$. A typical choice would be $\nu_\varepsilon \asymp \log(1/\varepsilon)$ or $\nu_\varepsilon \asymp \log\log(1/\varepsilon)$. Let

$$\rho_\varepsilon = \frac{1}{\log \nu_\varepsilon}.$$

Clearly, $\rho_\varepsilon \to 0$ as $\varepsilon \to 0$. Define the sequence $\{\kappa_j\}$ by

$$\kappa_j = \begin{cases} 1 & j = 0, \\ \nu_\varepsilon & j = 1, \\ \kappa_{j-1} + \lfloor \nu_\varepsilon \rho_\varepsilon (1 + \rho_\varepsilon)^{j-1} \rfloor & j = 2, \ldots, \end{cases} \tag{1.98}$$

where $\lfloor x \rfloor$ is the maximal integer that is strictly less than x. Let $\bar{N}$ be any integer satisfying

$$\bar{N} \geqslant \max \{ N : \varepsilon^2 \sum_{k=1}^{N} \sigma_k^2 \leqslant \rho_\varepsilon^{-3} \}. \tag{1.99}$$

Then, for ε small enough, $\bar{N} \geqslant \max \{ N : \varepsilon^2 \sum_{k=1}^{N} \sigma_k^2 \leqslant r^2 \rho_\varepsilon^{-2} \}$, $\forall r > 0$.

Remark 1.39. The term $\bar{N}$ is the final value of k. After that, the estimator is always fixed at 0. This $\bar{N}$ is fixed with the idea that the variance of a projection estimator $\hat{\theta}(N)$, i.e. $\varepsilon^2 \sum_{k=1}^{N} \sigma_k^2$, cannot be too large. Otherwise it is not even useful to consider larger values of N. Indeed, a good projection estimator should have a variance going to zero.

In this special construction assume the following:

(B1) *The blocks are* $I_j = [\kappa_{j-1}, \kappa_j - 1]$ *such that the values* κ_j *satisfy (1.98), and* $J = \min \{ j : \kappa_j > \bar{N} \}$ *where* $\bar{N}$ *satisfies (1.99).*
       Clearly, $N_{max} = k_J - 1 \geqslant \bar{N}$ if (B1) holds.
(B2) *The penalty is* $p_j = \Delta_{(j)}^a$, *where* $0 < a < 1/2$.

We also assume that the singular values b_k decrease precisely as a power of k:

(B3) *The coefficients b_k are positive and there exist $\beta \geqslant 0, b_* > 0$ such that*

$$b_k = b_* k^{-\beta}(1 + o(1)), \ k \to \infty.$$

Theorem 1.11. *Let θ^*_{pbs} be the penalized blockwise Stein estimator defined in (1.97). Assume (B1),(B2) and (B3), and let $r > 0$ be fixed. Then:*

(i) For any $\theta \in \ell^2$ such that $\|\theta\| \leqslant r$ and any $0 < \varepsilon < 1$ such that $\Delta_{(j)} \leqslant (1 - p_j)/4$ for all j, we have

$$\mathbf{E}_\theta \|\theta^*_{pbs} - \theta\|^2 \leqslant (1 + \tau_\varepsilon) \inf_{h \in \Lambda_{mon}} R(\theta, \lambda) + c\varepsilon^2 v_\varepsilon^{2\beta+1},$$

where $c > 0$ does not depend on θ, ε, and $\tau_\varepsilon = o(1)$, $\varepsilon \to 0$, τ_ε does not depend on θ.

(ii) For any $\lambda \in \Lambda_{mon}$ and $\theta \in \ell^2$ such that $R(\theta, \lambda) \leqslant r^2$ and any $0 < \varepsilon < 1$ such that $\Delta_{(j)} \leqslant (1 - p_j)/4$ for all j, we have

$$\mathbf{E}_\theta \|\theta^*_{pbs} - \theta\|^2 \leqslant (1 + \tau_\varepsilon)R(\theta, \lambda) + c\varepsilon^2 v_\varepsilon^{2\beta+1}.$$

Proof. A proof may be found in [32].

[44] consider their own block estimator, and show its sharp minimax adaptivity on the classes of ellipsoids.

A very long discussion, concerning, size of blocks, different penalties, several classes of functions where the estimator is minimax adaptive, may be found in [31]. The family of weakly geometrically increasing blocks is not in fact, the more precise choice in order to get very sharp results.

Other interesting results about the penalized Stein rule may be found in [14, 17] in the wavelet case and with heavy penalties p_j that do not tend to 0 as $T_j \to \infty$. In particular, [14] propose to take $p_j = 1/2 - 3/T_j$ and $T_j = 2^j$, while [17] considers small blocks with constant length $T_j \sim \log(1/\varepsilon)$ and $p_j > 4$. These penalties are too large to get exact oracle inequalities or sharp minimax adaptation, but they are sufficient for oracle inequality and then minimax adaptivity.

Remark 1.40. Since $\tau_\varepsilon = o(1)$, the oracle inequality of Theorem 1.11 may lead to some asymptotic exact oracle inequality. One needs to prove then that $\varepsilon^2 v_\varepsilon^{2\beta+1}$ is small.

In this part, we apply Theorem 1.11 to show that the penalized blockwise Stein estimator with the given special construction of blocks I_j is sharp minimax adaptive on the classes of ellipsoids.

Theorem 1.12. *Let $\Theta = \Theta(a, L)$ be an ellipsoid defined in (1.40) with monotone non-decreasing $a = \{a_k\}$, $a_k \to \infty$ and $L > 0$. Let the blocks I_j satisfy (B1), the penalties p_j satisfy (B2), and the singular values b_k satisfy (B3). Assume also that v_ε is chosen so that*

$$\frac{\varepsilon^2 v_\varepsilon^{2\beta+1}}{r_\varepsilon(\Theta)} = o(1), \ \varepsilon \to 0. \tag{1.100}$$

Then the penalized blockwise Stein estimator $\theta_{pbs}^\star = \{\theta_k^\star\}$ *defined in (1.97) is asymptotically minimax on* Θ *among all estimators, i.e.*

$$\sup_{\theta \in \Theta} \mathbf{E}_\theta \|\theta_{pbs}^\star - \theta\|^2 = r_\varepsilon(\Theta)(1 + o(1)), \tag{1.101}$$

as $\varepsilon \to 0$.

Proof. This is a simple consequence of Theorem 1.5 and Theorem 1.11. Note that under the assumptions of Theorem 1.12, the minimax sequence of Pinsker weights λ defined in (1.42) belongs to Λ_{mon}. Next, since a_k is monotone non-decreasing, $a_k \to \infty$, and b_k satisfies (B3), we have $r_\varepsilon(\Theta) \to 0$, as $\varepsilon \to 0$, by Theorem 1.5. Hence,

$$\sup_{\theta \in \Theta} R(\theta, \lambda) = r_\varepsilon^\ell(\Theta) = r_\varepsilon(\Theta)(1 + o(1)) = o(1),$$

as $\varepsilon \to 0$ where we used (1.45). Thus, the assumptions of Theorem 1.11 (ii) are satisfied for $\lambda = \lambda^p$ the Pinsker weights, $\theta \in \Theta$ and $r = 1$ if ε is small enough, and we may write

$$\sup_{\theta \in \Theta} \mathbf{E}_\theta \|\theta_{pbs}^\star - \theta\|^2 \leqslant (1 + o(1)) \sup_{\theta \in \Theta} R(\theta, \lambda^p) + c\varepsilon^2 v_\varepsilon^{2\beta+1}. \tag{1.102}$$

This, together with (1.100), yields

$$\sup_{\theta \in \Theta} \mathbf{E}_\theta \|\theta_{pbs}^\star - \theta\|^2 \leqslant r_\varepsilon^\ell(\Theta)(1 + o(1)),$$

which is equivalent to (1.101), in view of (1.45) and of the definition of $r_\varepsilon(\Theta)$.

Remark that Theorem 1.12 states the sharp adaptivity property of $\theta_{pbs}^\star$: this estimator is sharp asymptotically minimax on every ellipsoid $\Theta = \Theta(a, L)$ satisfying (1.100), while no prior knowledge about a and L is required to define $\theta_{pbs}^\star$.

Note also that the condition (1.100) is quite weak. It suffices to choose v_ε smaller than some iterated logarithm of $1/\varepsilon$, in order to satisfy these conditions for most of usual examples of ellipsoids Θ.

Corollary 1.2. *Let* $\Theta = \Theta(a, L)$ *be any ellipsoid with monotone non-decreasing* $a = \{a_k\}$ *such that* $k^{\alpha_1} \leqslant a_k \leqslant \exp(\alpha_2 k), \ \forall k,$ *for some* $\alpha_1 > 0, \alpha_2 > 0, L > 0$. *Assume (B1), (B2) and (B3) with* $v_\varepsilon = \max(\lfloor \log\log 1/\varepsilon \rfloor, 5)$. *Then the estimator* $\theta_{pbs}^\star$ *defined in (1.97) satisfies (1.101).*

Remark 1.41. The penalized blockwise Stein estimator is thus minimax adaptive on a very large scale of ellipsoids.

1.3.4.5 Model Selection Versus Universal Optimality

Comments

The approach of universal optimality, and the penalized blockwise Stein estimator, presented in Section 1.3.4 has very general and sharp properties.

Universal optimality. Theorem 1.11 shows that penalized blockwise Stein's estimator defined in (1.97) satisfies an oracle inequality on the class of all monotone sequences Λ_{mon}. In other words, it mimics the monotone oracle in (asymptotically) exact way. This immediately entails oracle inequalities on all the subclasses $\Lambda' \subset \Lambda_{mon}$. In particular, the estimator $\theta^\star_{pbs}$ is asymptotically at least as good as the optimal projection estimator, the optimal Tikhonov estimator or the optimal Pinsker estimator (see Section 1.2.2).
In a sense, this is a stronger property than oracle inequalities for the "model selection" estimators in Section 1.3.3 or [111, 82, 8, 25]. In those papers it was possible to treat in each occasion only one class Λ'.
This point is really crucial for the model selection approach. Among a family of estimators, one select the best possible one, by a data-driven selection method $\lambda^\star$ which takes its values in Λ.
The penalized blockwise Stein estimator at least mimics (and in fact outperforms) simultaneously the oracles on all these classes Λ'. This behaviour may be called *universal optimality*. One has then a universal estimator which is as good as most of the standard families of linear estimators.

Universal adaptivity. Another point is that, no "ellipsoidal" structure appears in the definition of $\theta^\star_{pbs}$. In fact, minimax results similar to Theorem 1.12 can be formulated for other classes than ellipsoids (for example, for parallelepipeds), provided the minimax solution λ is a monotone non-increasing sequence, see [31]. The penalized blockwise Stein estimator is thus minimax adaptive on a very large scale of classes of functions.
In a way, it is *universally adaptive.*

Non-linearity property. A last remark, is that the penalized blockwise Stein estimator is in fact, a non-linear estimator. Moreover, $\theta^\star_{pbs}$ does not even belong to the class Λ_{mon}.
It is well-known that on some classes of Besov classes with rather unsmooth functions, one needs non-linear estimators, for example wavelet thresholding, since linear ones are suboptimal, see [41]. [31] showed that $\theta^\star_{pbs}$ is (almost) optimal on these classes of unsmooth functions. The penalized blockwise Stein estimator, due to the shrinkage and the blocks, has, in some sense, the behaviour of a non-linear estimator.
Nevertheless, the penalized blockwise Stein estimator has some drawbacks.

Instability in inverse problems. The first one is almost the same than the URE estimator of Section 1.3.3.1. Due to the increase in the penalty the penalized version of blockwise Stein's estimator is less unstable than the URE estimator. However, one really needs a condition as a fixed N_{max} defined in (1.99) in order to avoid too large choices of blocks. Without this condition, the method is rather unstable in simulations.

Universal method: a constraint? A second drawback, of this universal approach is in fact its nature. Indeed, quite often in applications, scientists want to use their favourite method (Tikhonov, projection, $v-$method,...). They know, or believe, that this method works well in their field. In a way, the model selection approach answers to their problem. It allows to calibrate in a data-driven way (by choosing γ, N or m) their favourite method.

On the other hand, the universal approach, by its universal definition, does not really answer to their question. The universal optimality just proves that one very specific universal method, the penalized blockwise Stein estimator, is as good as their favourite method. However, this could be disappointing since they cannot use their own method.

Penalization in inverse problems. In a way, the penalized blockwise Stein estimator already contained the idea that in inverse problems, penalizing slightly more than the URE penalty was needed. Such a choice improves the accuracy of the method. The paper [31] was in fact written after [32]. In the inverse problems framework presented in [32] already appeared the need of stronger penalties than URE. This point is true in theory, but also in simulations where one has to be very careful with too large choices of number of coefficients N.

Nevertheless, after some times, the idea of penalizing slightly more than URE was found to be successful even in the context of the direct problem, i.e. Gaussian white noise. Thus, non-penalized blockwise Stein's rule leads to an oracle inequality which is similar, but less accurate than that of Theorem 1.11, see [31]. The study in [31] was also, in a way, deeper than in the inverse problems context. Indeed, there is a rather long discussion concerning, the different penalties, block sizes, and functional classes that may be studied.

The main idea was that one needed to penalize more than the URE penalty, especially in inverse problems. However, in order to get sharp theoretical results, but also a method accurate in simulations, this penalty did not need to be too large.

Thus, this was, in a way, the first step from unbiased risk estimation to risk hull method.

1.4 Conclusion

1.4.1 Summary

A very promising approach to inverse problems is the statistical framework. It is based on a model where observations contain a random noise. This does not correspond to the historical framework of [127] where the error is deterministic.

The optimal rates of convergence are different in the statistical and deterministic frameworks (see Section 1.2.5).

We have studied, in Section 1.1, the white noise model discretized in the spectral domain by use of the SVD, when the operator A is compact. This allows to define a measure of ill-posedness of an inverse problem, with influence on the rates of convergence.

Several examples of inverse problems where the SVD is known were presented (circular deconvolution, heat equation, tomography,...).

The spectral theory for non-compact operators was also developped with the example of deconvolution on $\mathbb{R}$.

In Section 1.2, the nonparametric approach and minimax point of view were presented. This notion corresponds to the asymptotic minimax optimality as the noise level goes to zero.

Several examples of standard regularization methods, and their counterpart as estimation procedures by use of SVD, were discussed (projection, Landweber, Tikhonov,...).

The notion of source condition was introduced, with its link with ellipsoid in ℓ^2 and standard classes of functions (Sobolev and analytic functions). The optimal rates of convergence were given. These rates depend on the smoothness of the function to reconstruct and on the degree of ill-posedness of the inverse problem.

In ill-posed inverse problems the rates are slower than in the direct problem, corresponding to the standard nonparametric statistics framework.

This notion of optimality leads to some optimal choice of the tuning parameter $(N, \gamma, \text{or } m)$.

However these optimal parameters are unachievable since they depend on the unknown smoothness of the function.

This remark leads to the main point of Section 1.3. The goal is to find data-driven choices of the tuning parameter (adaptive methods). In applications, this choice is just done by simulations in a very empirical way. For example, one uses known phantom images in order to calibrate, the estimator. Usually, there is no theoretical results in order to validate this approach. Moreover, this could be very unstable when the observed functions are different from the phantom.

The minimax adaptive approach is concerned with the construction of estimators which attain the optimal rates of convergence for any smoothness α of the function f.

The oracle approach is a second step in the problem of data-driven selection method. The oracle is the best possible choice, in a given family of estimators, provided we knew the unknown function. However, such a procedure cannot be constructed, since it is not an estimator. The aim of oracle inequalities is to prove that the estimator accuracy is close from the oracle behaviour.

There exist many different methods in order to construct data-driven choice of the tuning parameter. One of the more natural is the idea of minimizing an estimate of the risk (URE). The theoretical results concerning this method are satisfying.

Nevertheless, in simulations, the URE method is usually not stable enough in inverse problems. The approach of penalized empirical risk, may be better than URE provided the penalty function is chosen appropriately. The risk hull method (RHM) provides one way to find a good and explicit penalty function.

Another, adaptive method is considered, based on the blockwise Stein estimator. Again, with a slightly stronger penalty, this method is rather satisfying.

1.4.2 Discussion

The statistical approach to inverse problems is nowadays quite popular and successful.

There exist some differences between the two frameworks, stochastic and deterministic. For example, the optimal rates of convergence are not the same (see Section 1.2.5). Nevertheless, this difference in the optimal rates is not so important. In a way, the two frameworks are rather related.

However, one of the major advantages of the statistical approach is that it allows to obtain oracle inequalities and to construct adaptive estimators.

The oracle approach is thus very interesting in inverse problems. Indeed, one can construct procedures in order to choose the best estimator among a given family of regularization methods. This really gives some answer to a very natural problem, the data-driven choice of the tuning parameter (N or γ). From a practical point of view, this choice is usually just done by simulations in a very empirical way. Usually, by calibrating the method on some known phantom image. This approach may give a rather unstable procedure.

From a mathematical point of view, the oracle approach is very interesting. Indeed, the statistical theory is here able to give some answer to the very sensitive problem of data-driven choice of the tuning parameter.

Another important remark is that inverse problems are difficult problems. Indeed, we have to invert an operator in order to get the reconstruction. A main issue is then to get very precise oracle inequalities, i.e. with a good control on the constants of the main term, but also of the remainder term. The degree of ill-posedness of the problem appears in the rates of convergence, but at some point, in the oracle inequalities as well, which are thus sensitive to the difficulty of the problem.

Thus, in statistical inverse problems one has to define very precise model selection methods, or choice of the regularization parameter, otherwise, due to the difficulty of the problem, the estimator will not be accurate.

This remark is rather satisfying concerning the interest of the inverse problem framework in statistics. Indeed, due to the natural difficulty of the ill-posed problems, the statistical study is thus very challenging. In some sense, many estimators, or adaptive procedures, may be satisfying in the direct problem. Nevertheless, in the ill-posed context, one has to be much more careful, and the statistical study could really be more difficult.

1.4.3 Open Problems

In these lectures, the results have been obtained for a very specific and restrictive model. The model is a white noise model, with an additive and Gaussian noise. Moreover, a strong assumption is related to the use of SVD.

There exist many different approaches in order to extend the results or to deal with other kind of problems.

The goal of the present section is to discuss problems which are not presented in these notes. Several of these topics have been already well-studied in the literature, others remain more open.

Noisy Operators

One very restrictive assumption is that the operator A is perfectly known. Indeed, in many applications, the operator is not known, or at least not completely known. For example, in astronomical observations, point spread function may be changing due to unknown physical conditions. This problem is also related to the well-known problem of blind deconvolution, where one has to estimate also the convolution kernel.

From a theoretical point of view this problem is also quite challenging. Indeed, the operator, by its spectral behaviour, characterizes the optimal rates of convergence. Thus, it is not clear, if any modification on the operator would change the rates or not.

The case of fully unknown operator A is usually difficult. Indeed, one would need to estimate both the operator A and the function f by using the same data.

A more natural framework is the case of noisy operator, where the operator is not completely known and estimated using other data. This very important topic has been the subject of several recent statistical works, see for example in [48, 97, 66, 71]. In this framework, there exist two noises, one on the operator A and one on the inverse problem data. The main conclusion here is that, usually, the rate is the worse possible between these two noises.

A more specific model may also be considered, where the SVD basis is known, but the singular values are noisy. This setting appears for example in circular convolution model where the SVD is always the Fourier basis, but where the convolution kernel has to be estimated. In this situation, sharp oracle inequalities may be obtained, see [27, 29, 98].

Nondiagonal Case

One of the main drawbacks is that all these methods are linked to the spectral approach. We have intensively used the SVD to diagonalize the operators. The different regularization methods were presented for the spectral domain, even if, many of them can be computed without the explicit use of their SVD.

However, there is a more general situation where the operator A cannot be represented by a diagonal matrix. For example, one uses a basis, but which does not diagonalize the operator.

In this case, several results have been obtained, such that, optimal rates of convergence, adaptive estimation, oracle inequalities, see for example in [102, 97, 91].

Wavelets and Sparsity

As noted above, most of the methods are linked to the spectral approach. In many problems, this leads to the Fourier domain. Thus, due for example to the source conditions, the function to be reconstructed should have good properties in the Fourier domain.

Another very popular approach is based on wavelets, see for example in [42]. By using wavelets, one may usually deal with functions which are not very smooth, by replacing Sobolev classes by Besov classes. Indeed, there exist Besov classes which contain functions which are really unsmooth. Moreover, wavelets bases have the nice property that rather few coefficients are large, i.e. they give sparse representations. Thus, the standard estimator is constructed by using a threshold estimator of wavelets coefficients. This method allows to obtain adaptive estimators.

In inverse problems, wavelets have usually very good properties related to the operator A. Wavelets bases are not the exact SVD of a given operator. However, wavelets bases almost diagonalize many operators. Moreover by using thresholding they have good adaptability properties, see the Wavelet Vaguelette Decomposition (WVD) approach in [39]. This framework is thus strongly related to the previous nondiagonal case.

There exist a very large literature in inverse problems with wavelets, see for example in [39, 83, 78, 19, 28, 34, 79] and, with the framework of noisy operator, [71, 29].

RHM for Other Methods

The RHM is presented here for the family of projection estimators. There exist many other regularization methods (Landweber, Tikhonov, ν−methods). These methods usually attain the optimal rates of convergence, see for example [9]. The RHM approach has been very recently extended to these families of estimators (see [99]).

The RHM is also valid in the framework of noisy singular values, see [98].

Nonlinear Operators

All the results given here are valid for linear inverse problems. In the case of nonlinear operator, the problem is much more difficult. This framework has been intensively studied in the deterministic context, see [49].

However, this problem is not yet well understood in statistics. Due to the stochastic nature of the noise the nonlinear operator is more difficult to handle. Moreover, adaptive estimation and oracle inequalities are even more involved in this framework.

Some recent papers concerning the statistical study of nonlinear inverse problems may be found in [6, 92].

Error in Variables

There exist a rather popular topic in statistics which is very closely related to our framework, the error in variables problem. In this context, one observes

$$Y_i = X_i + \xi_i, \quad i = 1, \dots, n,$$

where $\{X_i\}$ and $\{\xi_i\}$ are i.i.d. random variables, independent, and usually defined on $\mathbb{R}$. The goal is to estimate the probability density f of the random variable X. Since X and ξ are independent, the probability density of Y is well-known to be a convolution of the two densities of X and ξ.

The exist two main differences here. The first one is that the model is a density type model, and not any more a white noise model. The second point is that the operator is usually not compact. Indeed, the convolution is on the whole $\mathbb{R}$ due to the random variables which take their values on $\mathbb{R}$. However, it is well-known that a white noise model may be considered as an idealized version of a density model, there even exists a formal equivalence, see [107]. Thus, usually the rates of convergence are the same in these two models, even if the mathematical proofs could be quite different.

Formally, this model could be considered as a special case of the model (1.2). However, the noise ξ should have a very specific behaviour, which is not true in the standard white noise case, see [9].

This model of error in variables is then really an inverse problem, and is often called the deconvolution problem in the statistical literature, see for example [52, 35, 16, 30].

Econometrics

Nowadays, the topic of inverse problems has also a growing interest in econometrics. The problem of intrumental variables is closely related to the framework of inverse problems.

An economic relationship between a response variable Y and a vector X of explanatory variables is often represented by the following equation

$$Y_i = f(X_i) + U_i, \quad i = 1, \dots, n,$$

where the function f has to be estimated and U_i are the errors. This model does not characterize the function f if U is not constrained. The problem is solved if $\mathbf{E}(U|X) = 0$.

However, in many structural econometrics models, the parameter of interest is a relation between Y and X, where some components of X are endogeneous. This situation arises frequently in economics. For example, suppose that Y denotes the wages and the X includes, level of education, among other variables. The error U includes, ability, which is not observed, but influences the wages. If a high ability tends to choose high level of education, then education and ability are correlated, and thus X and U also.

Nevertheless, suppose that we observe another set of data, W_i where W is called an instrumental variable for which

$$\mathbf{E}(U|W) = \mathbf{E}(Y - f(X)|W) = 0.$$

This equation characterizes f by a Fredholm equation of the first kind. Estimation of the function f is in fact an ill-posed inverse problems.

Since the years 2000, the framework of inverse problems has been the topic of many articles in the econometrics literature, see [54, 66, 33, 53], see also Chapter 2 by Jean-Pierre Florens in the present book.

Inverse Problems in Applications

These lectures mainly consider inverse problems from a theoretical point of view. This is satisfying from a mathematical perspective. Indeed, one can define this kind of problems with a rather general description. The difficulty of an inverse problem, i.e. its degree of ill-posedness β, is characterized by the spectral behaviour of the operator A when $k \to \infty$. Moreover, the smoothness α of the function f is also important. These parameters give the optimal rates of convergence.

This general description is rather important. Indeed, it allows to understand the difference between inverse problems, and the influence of the smoothness on the accuracy of the reconstruction.

However, all these concepts are mainly just mathematical tools. They are based on asymptotics, when k is large.

In a more applied point of view, there is mainly no difference between an inverse problem of degree $\beta = 2$ and a severely ill-posed problem. Moreover, many problems which are almost unsolvable are, in applications, rather easy to deal with. For example, a deconvolution by a Gaussian kernel ($\mathcal{N}(0, \sigma^2)$) is even worse that severely ill-posed. However, if this convolution kernel, has a small variance σ^2, then the problem is very easy to solve.

On the other hand, many of the problems which appear in applications are much more difficult than our framework with an idealized model. Even in the simple circular deconvolution, but with boundary effect, the SVD basis is not the Fourier basis anymore. The number of data in applications is finite n and does not go to infinity.

There is a lot of frameworks where identifiability of the model, i.e. existence and unicity, is the main problem, before any stability results. In more realistic models, most of the operators are not completely known and not even observed with some additive noise.

Tomography, even based on the same operator, the Radon transform, is a world by itself. There exist conferences and articles, on computerized tomography, positron emission tomography, discrete tomography, quantum homodyne tomography, see [104].

It is rather common to say, that each inverse problem is in fact a specific case, see [117].

Numerical Aspects

The numerical aspect of the different regularization methods was not so much discussed. However, this point is of importance, especially in inverse problems. As noted before, many of the regularization methods are expressed in the spectral domain (SVD) but many of them are in fact computed in a different way, without using the whole spectrum. For example, the Tikhonov regularization is computed by minimizing the functional (1.28). In deconvolution problems, the SVD will be the Fourier basis, which may really be computed quite fast by use of Fast Fourier Transform (FFT). In more difficult problems, for example Radon transform in tomography, the SVD could be much slower to compute, see [104, 49, 81, 130].

Moreover, iterative methods are rather popular because they also avoid the inversion of a large matrix as in (1.30). This is one the reason of the interest in all these iterative procedures, see [10].

Well-Posed Questions

One of the drawbacks of the study of inverse problems is its intrinsic difficulty. Indeed, the optimal rates of convergence may be rather slow (see Section 1.2.4). Even in the case of mildly ill-posed problems when the degree of ill-posedness β is large (even $\beta = 2$), the optimal rates will be quite slow. In the severely ill-posed context it is even worse and the rates could be logarithmic. Moreover, these rates are optimal, they cannot be improved on a given class of functions.

In a way some inverse problems are really too difficult (for example the heat equation). One may think that in a given model there is no hope to get better results.

A rather natural idea when a model is too difficult, is to change the goal of the problem. One tries to answer to problems that could be solved in a more satisfying way. The main point is thus to solve more easy problems than estimating the whole function f. For example, estimating linear functionals, level sets or change points, or solving testing or classification problems. It is well-known, that all these problems are more easy, i.e. have a better rate of convergence, than estimating the whole function f, see [56].

This point of view makes sense in problems where estimation of the whole function f seems almost beyond the scope. Thus, the idea is to find more simple tasks to deal with. In fact, one is looking to well-posed questions in ill-posed problems, see [117]. These are questions that may be answered in a satisfying way.

References

1. Adorf, H.M.: Hubble space telescope image restoration in its fourth year. Inverse Problems **11**, 639–653 (1995)
2. Akaike, H.: Information theory and an extension of the maximum likelihood principle. In: B. Petrov, F. Czáki (eds.) Proceedings of the Second International Symposium on Information Theory, pp. 267–281. Akademiai Kiadó, Budapest (1973)
3. Akaike, H.: A new look at the statistical model identification. IEEE Trans. Automat. Control **19**, 716–723 (1974)
4. Barron, A., Birgé, L., Massart, P.: Risk bounds for model selection via penalization. Probab. Theory Related Fields **113**, 301–413 (1999)
5. Bauer, F., Hohage, T.: A Lepski-type stopping rule for regularized Newton methods. Inverse Problems **21**, 1975–1991 (2005)
6. Bauer, F., Hohage, T., Munk, A.: Iteratively regularized Gauss-Newton method for nonlinear inverse problems with random noise. SIAM J. Numer. Anal. **47**, 1827–1846 (2009)
7. Belitser, E., Levit, B.: On minimax filtering on ellipsoids. Math. Methods Statist. **4**, 259–273 (1995)
8. Birgé, L., Massart, P.: Gaussian model selection. J. Eur. Math. Soc. **3**, 203–268 (2001)
9. Bissantz, N., Hohage, T., Munk, A., Ruymgaart, F.: Convergence rates of general regularizations methods for statistical inverse problems and application. SIAM J. Numer. Anal. **45**, 2610–2636 (2007)
10. Brakage, H.: On ill-posed problems and the method of conjuguate gradients. In: Inverse and ill-posed problems. Academic Press, Orlando (1987)
11. Bretagnolle, J., Huber, C.: Estimation des densités : risque minimax. Z. Wahrsch. Verw. Gebiete **47**, 199–237 (1976)
12. Brezis, H.: Analyse fonctionnelle, Théorie et applications. Dunod, Paris (1999)
13. Brown, L., Low, M.: Asymptotic equivalence of nonparametric regression and white noise. Ann. Statist. **24**, 2384–2398 (1996)
14. Brown, L., Low, M., Zhao, L.: Superefficiency in nonparametric function estimation. Ann. Statist. **25**, 898–924 (1997)
15. Bühlmann, P., Yu, B.: Boosting with ℓ^2-loss: regression and classification. J. Amer. Statist. Assoc. **98**, 324–339 (2003)
16. Butucea, C., Tsybakov, A.: Sharp optimality in density deconvolution with dominating bias. Theory Probab. Appl. **52**, 24–39 (2008)
17. Cai, T.: Adaptive wavelet estimation: a block thresholding and oracle inequality approach. Ann. Statist. **27**, 2607–2625 (1999)
18. Candès, E.: Modern statistical estimation via oracle inequalities. Acta Numer. **15**, 257–325 (2006)
19. Candès, E., Donoho, D.: Recovering edges in ill-posed inverse problems: Optimality of curvelet frames. Ann. Statist. **30**, 784–842 (2002)
20. Cavalier, L.: Efficient estimation of a density in a problem of tomography. Ann. Statist. **28**, 330–347 (2000)
21. Cavalier, L.: On the problem of local adaptive estimation in tomography. Bernoulli **7**, 63–78 (2001)
22. Cavalier, L.: Inverse problems with non-compact operator. J. of Statist. Plann. Inference **136**, 390–400 (2006)

23. Cavalier, L.: Nonparametric statistical inverse problems. Inverse Problems **24**, 1–19 (2008)
24. Cavalier, L., Golubev, G., Lepski, O., Tsybakov, A.: Block thresholding and sharp adaptive estimation in severely ill-posed inverse problems. Theory Probab. Appl. **48**, 426–446 (2003)
25. Cavalier, L., Golubev, G., Picard, D., Tsybakov, A.: Oracle inequalities in inverse problems. Ann. Statist. **30**, 843–874 (2002)
26. Cavalier, L., Golubev, Y.: Risk hull method and regularization by projections of ill-posed inverse problems. Ann. Statist. **34**, 1653–1677 (2006)
27. Cavalier, L., Hengartner, N.: Adaptive estimation for inverse problems with noisy operators. Inverse Problems **21**, 1345–1361 (2005)
28. Cavalier, L., Koo, J.Y.: Poisson intensity estimation for tomographic data using a wavelet shrinkage approach. IEEE Trans. Inform. Theory **48**, 2794–2802 (2002)
29. Cavalier, L., Raimondo, M.: Wavelet deconvolution with noisy eigenvalues. IEEE Trans. Signal Process. **55**, 2414–2424 (2007)
30. Cavalier, L., Raimondo, M.: Multiscale density estimation with errors in variables. J. Korean Statist. Soc. (2010)
31. Cavalier, L., Tsybakov, A.: Penalized blockwise Stein's method, monotone oracles and sharp adaptive estimation. Math. Methods Statist. **10**, 247–282 (2001)
32. Cavalier, L., Tsybakov, A.: Sharp adaptation for inverse problems with random noise. Probab. Theory Related Fields **123**, 323–354 (2002)
33. Chen, X., Reiss, M.: On rate optimality for ill-posed inverse problems in econometrics (2010). In press
34. Cohen, A., Hoffmann, M., Reiss, M.: Adaptive wavelet Galerkin method for linear inverse problems. SIAM J. Numer. Anal. **42**, 1479–1501 (2004)
35. Comte, F., Rozenholc, Y., Taupin, M.L.: Penalized contrast estimator for adaptive density deconvolution. Canad. J. Statist. **34**, 431–452 (2006)
36. Craven, P., Wahba, G.: Smoothing noisy data with spline functions: estimating the correct degree of smoothing by the method of generalized cross-validation. Numer. Math. **31**, 377–403 (1979)
37. Deans, S.: The Radon Transform and some of its Applications. Wiley, New York (1983)
38. Donoho, D.: Statistical estimation and optimal recovery. Ann. Statist. **22**, 238–270 (1994)
39. Donoho, D.: Nonlinear solution of linear inverse problems by wavelet-vaguelette decomposition. Appl. Comput. Harmon. Anal. **2**, 101–126 (1995)
40. Donoho, D., Johnstone, I.: Ideal spatial adaptation via wavelet shrinkage. Biometrika **81**, 425–445 (1994)
41. Donoho, D., Johnstone, I.: Adapting to unknown smoothness via wavelet shrinkage. J. Amer. Statist. Assoc. **90**, 1200–1224 (1995)
42. Donoho, D., Johnstone, I.: Minimax estimation via wavelet shrinkage. Ann. Statist. **26**, 879–921 (1998)
43. Donoho, D., Low, M.: Renormalization exponents and optimal pointwise rates of convergence. Ann. Statist. **20**, 944–970 (1992)
44. Efroimovich, S., Pinsker, M.: Learning algorithm for nonparametric filtering. Autom. Remote Control **11**, 1434–1440 (1984)
45. Efromovich, S.: Robust and efficient recovery of a signal passed through a filter and then contaminated by non-Gaussian noise. IEEE Trans. Inform. Theory **43**, 1184–1191 (1997)
46. Efromovich, S.: Nonparametric Curve Estimation. Springer, New York (1998)
47. Efromovich, S.: Simultaneous sharp adaptive estimation of functions and their derivatives. Ann. Statist. **26**, 273–278 (1998)
48. Efromovich, S., Koltchinskii, V.: On inverse problems with unknown operators. IEEE Trans. Inform. Theory **47**, 2876–2893 (2001)
49. Engl, H., Hanke, M., Neubauer, A.: Regularization of Inverse Problems. Kluwer Academic Publishers (1996)
50. Ermakov, M.: Minimax estimation of the solution of an ill-posed convolution type problem. Problems Inform. Transmission **25**, 191–200 (1989)
51. Evans, S., Stark, P.: Inverse problems as statistics. Inverse Problems **18**, 55–97 (2002)

52. Fan, J.: On the optimal rates of convergence for nonparametric deconvolution problems. Ann. Statist. **19**, 1257–1272 (1991)
53. Florens, J., Johannes, J., Van Bellegem, S.: Identification and estimation by penalization in nonparametric instrumental regression. Econ. Theory (2010). In press
54. Florens, J.P.: Inverse problems and structural econometrics: the example of instrumental variables. In: Advances in Economics and Econometrics: Theory and Applications, vol. 2, pp. 284–311 (2003)
55. Goldenshluger, A.: On pointwise adaptive nonparametric deconvolution. Bernoulli **5**, 907–925 (1999)
56. Goldenshluger, A., Pereverzev, S.: Adaptive estimation of linear functionals in Hilbert scales from indirect white noise observations. Probab. Theory Related Fields **118**, 169–186 (2000)
57. Goldenshluger, A., Spokoiny, V.: On the shape-from-moments problem and recovering edges from noisy Radon data. Probab. Theory Related Fields **128**, 123–140 (2004)
58. Goldenshluger, A., Tsybakov, A.: Adaptive prediction and estimation in linear regression with infinitely many parameters. Ann. Statist. **29**, 1601–1619 (2001)
59. Golubev, G.: Quasi-linear estimates of signals in L^2. Problems Inform. Transmission **26**, 15–20 (1990)
60. Golubev, G.: The principle of penalized empirical risk in severely ill-posed problems. Probab. Theory Related Fields **130**, 18–38 (2004)
61. Golubev, G., Khasminskii, R.: A statistical approach to some inverse problems for partial differential equations. Problems Inform. Transmission **35**, 51–66 (1999)
62. Golubev, G., Khasminskii, R.: A statistical approach to the Cauchy problem for the Laplace equation. Lecture Notes Monograph Series **36**, 419–433 (2001)
63. Grama, I., Nussbaum, M.: Asymptotic equivalence for nonparametric regression. Math. Methods Statist. **11**, 1–36 (2002)
64. Groetsch, C.: Generalized Inverses of Linear Operators: Representation and Approximation. Dekker, New York (1977)
65. Hadamard, J.: Le problème de Cauchy et les équations aux dérivées partielles hyperboliques. Hermann, Paris (1932)
66. Hall, P., Horowitz, J.: Nonparametric methods for inference in the presence of instrumental variables. Ann. Statist. **33**, 2904–2929 (2005)
67. Hall, P., Kerkyacharian, G., Picard, D.: Block threshold rules for curve estimation using kernel and wavelet methods. Ann. Statist. **26**, 922–942 (1998)
68. Halmos, P.: What does the spectral theorem say? Amer. Math. Monthly **70**, 241–247 (1963)
69. Hida, T.: Brownian Motion. Springer-Verlarg, New York-Berlin (1980)
70. Hoerl, A.: Application of ridge analysis to regression problems. Chem. Eng. Progress **58**, 54–59 (1962)
71. Hoffmann, M., Reiss, M.: Nonlinear estimation for linear inverse problems with error in the operator. Ann. Statist. **36**, 310–336 (2008)
72. Hohage, T.: Lecture notes on inverse problems (2002). Lectures given at the University of Göttingen
73. Hutson, V., Pym, J.: Applications of Functional Analysis and Operator Theory. Academic Press, London (1980)
74. Ibragimov, I., Khasminskii, R.: Statistical Estimation: Asymptotic Theory. Springer, New York (1981)
75. Ibragimov, I., Khasminskii, R.: On nonparametric estimation of the value of a linear functional in Gaussian white noise. Theory Probab. Appl. **29**, 19–32 (1984)
76. James, W., Stein, C.: Estimation with quadratic loss. In: Proceedings of the 4th Berkeley Symposium on Mathematical Statistics and Probability, pp. 361–380. University of California Press (1961)
77. Johnstone, I.: Function estimation in Gaussian noise: sequence models (1998). Draft of a monograph
78. Johnstone, I.: Wavelet shrinkage for correlated data and inverse problems: adaptivity results. Statist. Sinica **9**, 51–83 (1999)

79. Johnstone, I., Kerkyacharian, G., Picard, D., Raimondo, M.: Wavelet deconvolution in a periodic setting. J. R. Stat. Soc. Ser. B Stat. Methodol. **66**, 547–573 (2004)
80. Johnstone, I., Silverman, B.: Speed of estimation in positron emission tomography and related inverse problems. Ann. Statist. **18**, 251–280 (1990)
81. Kaipio, J., Somersalo, E.: Statistical and Computational Inverse Problems. Springer (2004)
82. Kneip, A.: Ordered linear smoothers. Ann. Statist. **22**, 835–866 (1994)
83. Kolaczyk, E.: A wavelet shrinkage approach to tomographic image reconstruction. J. Amer. Statist. Assoc. **91**, 1079–1090 (1996)
84. Koo, J.Y.: Optimal rates of convergence for nonparametric statistical inverse problems. Ann. Statist. **21**, 590–599 (1993)
85. Korostelev, A., Tsybakov, A.: Optimal rates of convergence of estimators in a probabilistic setup of tomography problem. Probl. Inf. Transm. **27**, 73–81 (1991)
86. Landweber, L.: An iteration formula for Fredholm equations of the first kind. Amer. J. Math. **73**, 615–624 (1951)
87. Lepskii, O.: One problem of adaptive estimation in Gaussian white noise. Theory Probab. Appl. **35**, 459–470 (1990)
88. Lepskii, O.: Asymptotic minimax adaptive estimation. 1. Upper bounds. Theory Probab. Appl. **36**, 654–659 (1991)
89. Lepskii, O.: Asymptotic minimax adaptive estimation. 2. Statistical model without optimal adaptation. Adaptive estimators. Theory Probab. Appl. **37**, 468–481 (1992)
90. Li, K.C.: Asymptotic optimality of C_P, C_L, cross-validation and generalized cross-validation: Discrete index set. Ann. Statist. **15**, 958–976 (1987)
91. Loubes, J.M., Ludena, C.: Adaptive complexity regularization for linear inverse problems. Electron. J. Stat. **2**, 661–677 (2008)
92. Loubes, J.M., Ludena, C.: Penalized estimators for nonlinear inverse problems. ESAIM Probab. Stat. (2010)
93. Loubes, J.M., Rivoirard, V.: Review of rates of convergence and regularity conditions for inverse problems. Int. J. Tomogr. and Stat. (2009)
94. Louis, A., Maass, P.: A mollifier method for linear operator equations of the first kind. Inverse Problems **6**, 427–440 (1990)
95. Mair, B., Ruymgaart, F.: Statistical estimation in Hilbert scale. SIAM J. Appl. Math. **56**, 1424–1444 (1996)
96. Mallows, C.: Some comments on C_p. Technometrics **15**, 661–675 (1973)
97. Marteau, C.: Regularization of inverse problems with unknown operator. Math. Methods Statist. **15**, 415–443 (2006)
98. Marteau, C.: On the stability of the risk hull method for projection estimators. J. Statist. Plann. Inference **139**, 1821–1835 (2009)
99. Marteau, C.: Risk hull method for general family of estimators. ESAIM Probab. Stat. (2010)
100. Massart, P.: Concentration Inequalities and Model Selection. Lecture Notes in Mathematics, Springer, Berlin (2007)
101. Mathé, P.: The Lepskii principle revisited. Inverse Problems **22**, 11–15 (2006)
102. Mathé, P., Pereverzev, S.: Optimal discretization of inverse problems in Hilbert scales. Regularization and self-regularization of projection methods. SIAM J. Numer. Anal. **38**, 1999–2021 (2001)
103. Mathé, P., Pereverzev, S.: Regularization of some linear ill-posed problems with discretized random noisy data. Math. Comp. **75**, 1913–1929 (2006)
104. Natterer, F.: The Mathematics of Computerized Tomography. J. Wiley, Chichester (1986)
105. Nemirovski, A.: Topics in Non-Parametric Statistics. Lecture Notes in Mathematics, Springer (2000)
106. Nemirovskii, A., Polyak, B.: Iterative methods for solving linear ill-posed problems under precise information I. Engrg. Cybernetics **22**, 1–11 (1984)
107. Nussbaum, M.: Asymptotic equivalence of density estimation and Gaussian white noise. Ann. Statist. **24**, 2399–2430 (1996)
108. O'Sullivan, F.: A statistical perspective on ill-posed problems. Statist. Sci. **1**, 502–527 (1986)

109. Pinsker, M.: Optimal filtering of square integrable signals in Gaussian white noise. Problems Inform. Transmission **16**, 120–133 (1980)
110. Plaskota, L.: Noisy Information and Computational Complexity. Cambridge University Press (1996)
111. Polyak, B., Tsybakov, A.: Asymptotic optimality of the C_p-test for the orthogonal series estimation of regression. Theory Probab. Appl. **35**, 293–306 (1990)
112. Raus, T., Hamarik, U., Palm, R.: Use of extrapolation in regularization methods. J. Inverse Ill-Posed Probl. **15**, 277 (2007)
113. Reiss, M.: Asymptotic equivalence for nonparametric regression with multivariate and random design. Ann. Statist. **36**, 1957–1982 (2008)
114. Rigollet, P.: Adaptive density estimation using the blockwise Stein method. Bernoulli **12**, 351–370 (2006)
115. Rosenblatt, M.: Remarks on some nonparametric estimates of a density function. Ann. Math. Statist. **27**, 832–837 (1956)
116. Ruymgaart, F.: A short introduction to inverse statistical inference (2001). Lecture given at Institut Henri Poincaré, Paris
117. Sabatier, P.: Past and future of inverse problems. J. Math. Phys. **41**, 4082–4124 (2000)
118. Schwarz, G.: Estimating the dimension of a model. Ann. Statist. **6**, 461–464 (1978)
119. Shibata, R.: An optimal selection of regression variables. Biometrika **68**, 45–54 (1981)
120. Stein, C.: Inadmissibility of the usual estimator of the mean of a multivariate distribution. In: Proceedings of the 3rd Berkeley Symposium on Mathematical Statistics and Probability, pp. 197–206. University of California Press (1956)
121. Stein, C.: Estimation of the mean of a multivariate normal distribution. Ann. Statist. **9**, 1135–1151 (1981)
122. Stone, C.: Optimal rates of convergence for nonparametric estimators. Ann. Statist. **8**, 1348–1360 (1980)
123. Sudakov, V., Khalfin, L.: Statistical approach to ill-posed problems in mathematical physics. Soviet Math. Dokl. **157**, 1094–1096 (1964)
124. Talagrand, M.: Concentration of measure and isoperimetric inequalities in product spaces. Publ. Math. IHES **81**, 73–205 (1995)
125. Taylor, M.: Partial differential equations, vol. 2. Springer, New York (1996)
126. Tenorio, L.: Statistical regularization of inverse problems. SIAM Rev. **43**, 347–366 (2001)
127. Tikhonov, A.: Regularization of incorrectly posed problems. Soviet Math. Dokl. **4**, 1624–1627 (1963)
128. Tikhonov, A., Arsenin, V.: Solution of Ill-posed Problems. Winston & Sons (1977)
129. Tsybakov, A.: Introduction to Nonparametric Estimation. Springer series in statistics (2009)
130. Vogel, C.: Computational Methods for Inverse Problems. SIAM, Philadelphia (2002)
131. Wahba, G.: Spline Models for Observational Data. SIAM, Philadelphia (1990)

Part II
Invited Contribution on Inverse Problems

Chapter 2
Non-parametric Models with Instrumental Variables

Jean-Pierre Florens

Abstract This chapter gives a survey of econometric models characterized by a relation between observable and unobservable random elements where these unobservable terms are assumed to be independent of another set of observable variables called instrumental variables. This kind of specification is usefull to address the question of endogeneity or of selection bias for examples. These models are treated non parametrically and, in all the examples we consider, the functional parameter of interest is defined as the solution of a linear or non linear integral equation. The estimation procedure then requircs to solve a (generally ill-posed) inverse problem. We illustrate the main questions (construction of the equation, identification, numerical solution, asymptotic properties, selection of the regularization parameter) with the different models we present.

2.1 Introduction

Most of the econometric models take the form of a relation between a random element Y and two others random elements Z and U. Both Y and Z are observable (we have for example an *i.i.d.* sample $(y_i, z_i)_{i=1,...,n}$ of (Y, Z)) but U is unobservable. In econometrics U may be view as a summary of all the missing variables of the model. The form of the relation may vary. Consider for example the three following cases:

i) $Y = \langle Z, \varphi \rangle + U$ where $\langle Z, \varphi \rangle$ denotes a scalar product between Z and a parameter φ (Z and φ may be infinite dimensional)

ii) $Y = \varphi(Z) + U$ where φ is an unknown function of Z.

iii) $Y = \varphi(Z, U)$ where φ is an unknown function of Z and U and is assumed to be increasing w.r.t. U. In this model the distribution of U is given.

Jean-Pierre Florens
Université Toulouse 1 & Toulouse School of Economics, GREMAQ & IDEI, 21 allée de Brienne, 31000 Toulouse, France, e-mail: florens@cict.fr

The two first cases are said separable and the last one is non separable. We will say that Z is exogenous if the object of interest (the function φ) is characterized by an "independence-type" between Z and U. In the first case this condition can be relaxed to a non correlation condition $E(ZU) = 0$ and $\langle Z, \varphi \rangle$ is the linear regression of Y relatively to Z. In the second case a mean "independence-type" $E(U|Z) = 0$ is enough and φ is equal to the conditional expectation of Y given Z. In the last case it is usually assumed that U and Z are fully independent. If moreover U is uniformly distributed between 0 and 1, $\varphi(Z, U)$ is the quantile function of Y given Z. At least in the two last cases, the exogeneity condition means that φ is determined by the conditional distribution of Y given Z.

As economics is not in general an experimental science, the exogeneity assumption creates an analogous statistical framework to treat economic data as in an experimental context. Essentially the econometrician may treat the observations of Z as if they were fixed by an experimentalist and the mechanism generating the Z may be neglected in the estimation process of φ. This concept of exogeneity is fundamental in econometrics and has been analyzed from the beginning of econometric's researchs (see [32]) or more recently in connection to the concept of cut in statistical model (see [11], [18]).

However in many important applications of statistics to economic data an exogeneity assumption is not valid in the sense that it does not characterizes the parameter of interest. The elementary following example illustrates this point : assume Y and Z are real and we are interested in the parameter φ of a linear relation $Y = \varphi Z + U$. The variable Z is generated according to an equation $Z = \gamma W + V$ where W is an observable variable and V is an unobservable noise correlated with U. In that case $E(ZU) \neq 0$. There exists a parameter β such that $E((Y - \beta Z)Z) = 0$ but this parameter is different from φ.

We say that Z is endogenous if Z is not exogenous. This definition is not operational and should be made more precise in order to lead to a characterization of the parameter of interest.

The endogeneity of the Z variable can be illustrated in the case of of treatment effects which is not specific to econometrics but which is very useful to motivate the interest to endogenous variables. Consider for example a deterministic variable $\zeta \in R$ representing the level of a treatment and Y is a random element denoting the outcome of the treatment. Let us assume that the impact of the treatment ζ on Y may be formalized by a relation $Y = \varphi(\zeta) + U$ where $\varphi(\zeta)$ represents the average effect of a level of treatment equal to ζ (i.e. $E(U) = 0$). In a non experimental design the level of the treatment Z assigned to an individual is not randomly determined but may depend on some characteristics of the patient observable by the person who fix the treatment but not by the statistician. In that case the model used by the statistician is $Y = \varphi(Z) + U$ but the assumption $E(U|Z) = 0$ is not relevant.

This example may be extended to macro econometric analysis. The aggregated consumption of some good may be written $Y = \varphi(\pi) + U$ where π is a fixed non random value of the price of this good. The function φ is in that case the averaged aggregated demand function. The observed price P is not at all randomly generated and follows for example from the equilibrium of a system of demand and supply (the supply verifies $S = \psi(\pi) + V$ and the statistician observes Y and P such that $Y = S$ or $\varphi(P) + U = \psi(P) + V$). In this situation the model becomes $Y = \varphi(P) + U$ but $E(U|P) \neq 0$.

In most of the cases Z is endogenous because it is not fixed or randomized but is generated including a strategic component of the economic agents or Z follows from an equilibrium rule among the economic agents.

The three models we have introduced before are not well defined if we eliminate the independence assumption between Z and U. These assumptions should be replace by other assumptions in order to characterize the function φ.

The more natural extension to models with exogenous variables is provided by models with instrumental variables (IV). We consider now three random elements (Y, Z, W) where W are the instruments and the model is still specified by a relation linking Y to Z and U but U is now assumed to verify an "independence-type" property with W and not with Z. This approach extends obviously the exogeneity case because W and Z may be taken equal. But the interest of this framework is to separate the relevant variables in the model (Z) and the variables independent to the residual (W). In a general presentation Z and W may have common elements but contain specific variables. In the three models presented above the "independence-type" are indeed $E(WU) = 0, E(U|W) = 0$ or $U \perp\!\!\!\perp W$ (U and W independent).

The IV approach is not the unique way to formalize the endogeneity condition. In separable models, we may introduce a control function approach. Consider for example the second model and let us compute the conditional expectation $E(Y|Z, W) = \varphi(Z) + E(U|Z, W)$. A control function approach is based on the assumption that there exist a function $C(W, Z)$ such that $E(U|Z, W) = E(U|C) = \psi(C)$ and such that C is sufficiently separated of Z to allow the identification of the two components of the additive model $Y = \varphi(Z) + \psi(C) + \varepsilon$. For example we may assumed that $\frac{\partial}{\partial Z} \psi(C) = 0$. In that case $\varphi(Z)$ is obtained up to an additive constant by solving the equation $E(Y|Z, C) = \varphi(Z) + \psi(C)$ (see [35] or [15]).

In this chapter we focus our attention on the instrumental variables approach in a non parametric context. This question has generated numerous researches in the last ten years in econometrics and this chapter is just a survey of the main elements of this literature. The goal is to present the key points through different examples.

The strategy to examine this question is the following. First we derive from the "independence-type" between U and W a functional equation which links the unknown object of interest φ and the probability distribution of (Y, Z, W). Under the hypothesis of correct specification we assume that a solution of this equation exists. The second question is the uniqueness or local uniqueness of this solution, or, in econometric terminology, the question of identification or local identification. This uniqueness property usually requires some dependence condition between the Z and the W variables. In the third step, we use the equation derived from the "independence-type" between U and W to estimate φ. We replace the distribution of (Y, Z, W) by a non parametric estimate and we estimate φ as the solution of the estimated equation. Unfortunately this simple approach based on the resolution of an estimated functional equation belongs in general to the class of ill-posed inverse problems and this naïve solution is not a consistent estimator. This difficulty is solved by a penalization technique and we essentially consider in this chapter L^2 penalizations. The final element consists in examining the asymptotic properties of the estimator and in deriving in particular its rate of convergence to the true function. This rate will basically depend upon the difficulty of the resolution of the equation ("degree of ill-posedness") and of the regularity of φ relative to the problem ("degree of smoothness"). In the IV case the degree of ill-posedness is related to the dependence between the Z and the W. Intuitively low dependence means high degree of ill-posedness.

The realization of the penalized problem of the equation requires the choice of some regularization parameter and a data driven selection of this parameter is essential for the implementation of this approach. A comparison between a feasible estimator based on the data driven selection of the regularization parameter and a theoretical unfeasible estimator based on an optimal selection of the regularization parameter is important and may be conducted in the spirit of "oracle" inequalities." This last point will not be treated in this chapter (see Chapter 1 by Laurent Cavalier in this volume or in a Bayesian context [14]).

This chapter will review the instrumental variable analysis in the three models introduced. We also briefly introduce the extension of the previous ideas to some dynamic models, essentially developed a in static framework.

2.2 The Linear Model: Vectorial or Functional Data

Let us start to recall the elementary model of instrumental variables which reduces to the well known two stages least squares method in the homoscedastic case.

We consider a random vector (Y, Z, W) where $Y \in R, Z \in R^p$ and $W \in R^q$ (Z and W may have common elements) and the model verifies:

$$\begin{cases} Y = Z'\beta + U \\ E(WU) = 0 \end{cases} \qquad (2.1)$$

where $\beta \in R^p$ is the parameter of interest.

The condition 2.1 leads to the equation:

$$E(WZ')\beta = E(WY) \qquad (2.2)$$

denoted $T\beta = r$ with T is a $q \times p$ - matrix operator from R^p to R^q and r is an element of R^q. This system of linear equations is assumed to have a solution (well specification of the model) and this solution is unique (identification condition) if T is one to one, i.e. if $E(ZW')$ has a rank equal to p (which needs in particular $q \geqslant p$). This system is solved through the minimization of

$$\|T\beta - r\|^2 \qquad (2.3)$$

where the norm is the euclidian norm in R^q and the solution is

$$\beta = (T'T)^{-1}T'r \tag{2.4}$$

where T' denotes the transpose of T.

We assume that an *i.i.d.* sample $(y_i, z_i, w_i)_{i=1,\dots,n}$ is available and the estimation of β is obtained by the replacement of T, T' and r by their empirical counterparts:

$$\hat{\beta} = \left[\left(\frac{1}{n}\sum_{i=1}^{n} z_i w_i'\right)\left(\frac{1}{n}\sum_{i=1}^{n} w_i z_i'\right)\right]^{-1}\left(\frac{1}{n}\sum z_i w_i'\right)\left(\frac{1}{n}\sum w_i y_i\right) \tag{2.5}$$

This estimator is not optimal in terms of its asymptotic variance. To find an optimal estimator we may start again from the moment condition $E(W(Y - Z'\beta)) = 0$ and the usual results (see [26]) on GMM (Generalized Moments Method) implies that optimal estimation is deduced from the minimization of

$$\|B(T\varphi - r)\|^2 \text{ where } B = [Var(WU)]^{-\frac{1}{2}} \tag{2.6}$$

This minimization gives

$$\beta = (T'B'BT)^{-1}T'B'Br \tag{2.7}$$

If $Var(U|W) = \sigma^2$ (homoscedastic case) $B'B$ reduces to $[Var(W)]^{-1}$ or, using the empirical counter parts of these operators, we have the usual two stages estimator:

$$\hat{\beta} = \left[\left(\frac{1}{n}\sum_{i=1}^{n} z_i w_i\right)\left[\left(\frac{1}{n}\sum_{i=1}^{n} w_i w_i'\right)\right]^{-1}\left(\frac{1}{n}\sum_{i=1}^{n} w_i z_i'\right)\right]^{-1}$$
$$\frac{1}{n}\sum_{i=1}^{n} z_i w_i \left(\frac{1}{n}\sum_{i=1}^{n} w_i w_i'\right)^{-1}\frac{1}{n}\sum_{i=1}^{n} w_i y_i \tag{2.8}$$

This estimation is consistent and verifies

$$\sqrt{n}(\hat{\beta} - \beta) \Rightarrow N(0, \sigma^2(T'B'BT)^{-1}) \tag{2.9}$$

This computation requires the inversion of two matrix and it is natural to consider questions coming from the possible ill conditioning of these matrix. The inversion of $Var(W)$ may be difficult if the dimension of W becomes large, in particular if the sample size is small compared to q the dimension of W. This difficulty may be solved by a regularization of the inversion of this variance and $\frac{1}{n}\Sigma w_i w_i'$ may be replaced by $\alpha I + \frac{1}{n}\Sigma w_i w_i'$ where α is a positive parameter going to 0 when $N \to \infty$ (see [5], [4]).

Another question comes from the rank condition on $E(WZ')$ which determines the identification condition. A recent literature on the so called "weak instruments" (see a survey by [36]) considers cases where rank $(\frac{1}{n}\Sigma_{i=1}^{n} w_i z_i') = p$ but where this matrix converges to a non full rank matrix. The correct mathematical formalization of this situation is not very easy if the dimension of the vector W and Z are kept fixed. This question is more easy to understand in the case where the dimension of Z and W are infinite.

The natural extension of the previous model (see [22]) considers $Y \in R, Z \in \mathscr{F}$ and $W \in \mathscr{H}$ where $\mathscr{F}$ and $\mathscr{H}$ are two Hilbert spaces.

The model now becomes:

$$Y = \langle Z, \varphi \rangle + U \quad \varphi \in \mathscr{F} \langle, \rangle \text{ scalar product in } \mathscr{F} \tag{2.10}$$

$$E(WU) = 0 \tag{2.11}$$

where φ is the functional parameter of interest and where the condition 2.11 involves an expectation in the space $\mathscr{H}$. For example if $\mathscr{F}$ is the L^2 space of square integrable functions defined on $[0,1]$ with respect to the Lebesgue measure we have

$$\langle Z, \varphi \rangle = \int_0^1 Z(t)\varphi(t)dt \tag{2.12}$$

or if $\mathscr{F}$ is ℓ^2 space of square sommable sequences with respect to a measure $(\pi_j)_{j=0,1\ldots}$ we may have

$$\langle Z, \varphi \rangle = \sum_{j=0}^{\infty} Z_j \varphi_j \pi_j \tag{2.13}$$

The functional equation determined by condition 2.10 is now rewritten

$$E(W\langle Z, \varphi \rangle) = E(WY) \tag{2.14}$$

or $T\varphi = r$ where T is the covariance operator from $\mathscr{F}$ to $\mathscr{H}$. We still assume the model well specified (a solution exists to 2.13) and identified (T is one to one).

The equation $T\varphi = r$ which characterizes φ is now a "Fredholm equation of type I" and is ill-posed when the covariance operator T is compact. In that case the generalized inverse solution 2.4 is not a continuous function of r and then does not lead to a consistent estimator.

The resolution of $T\varphi = r$ is then an ill-posed linear inverse problem which has the particularity that not only r is estimated but that the operator T is also unknown and estimated using the same data set as r.

The estimation of φ will be performed using a regularization technique and we will concentrate here on the estimation by a Tikhonov regularization which may include a smoothness constraint (see [25] for the case where Z is exogenous).

Let $L : \mathscr{F} \to \mathscr{F}$ a differential operator defined on a dense subset of $\mathscr{F}$ and self adjoint. For example let us take the operator I on $L^2[0,1]$ defined by:

$$I\varphi = \int_0^t \varphi(s)ds \tag{2.15}$$

and let us define L by $L^{-2} = I^*I$. The $*$ denotes the adjoint operator. We easily see that $\varphi \in \mathscr{D}(L^{-b})$ is equivalent to say that φ is b differentiable and satisfies some boundary conditions (e.g. in our example $\varphi \in \mathcal{D}(L^{-2})$ means that φ is twice differentiable and $\varphi(0) = \varphi'(0) = 0$).

Let us assume that $\varphi \in \mathcal{D}(L^{-b})$ an consider $s \leqslant b$.

We consider the following Tikhonov functional

$$\|T\varphi + r\|^2 + \alpha\|L^s\varphi\|^2 \tag{2.16}$$

The minimum φ^α is equal to:

$$\varphi^\alpha = (\alpha L^{2s} + T^*T)^{-1}T^*r$$
$$= L^{-s}(\alpha I + L^{-s}T^*TL^{-s})^{-1}L^{-s}T^*r \tag{2.17}$$

and the estimator is obtained by replacing T, T^* and r by their empirical counterparts $\hat{T}, \hat{T}^*$ and $\hat{r}$.

At least three questions follows from this estimation mechanism: is the estimator easily computable, what are its asymptotic properties in relation in particular to a concept of strong or weak instruments and is it possible to extend the optimality argument presented in the finite dimensional case to the functional linear model. The answers of these question are given in [22] and we just summarize here the main results.

Let us first remark that the computation of the estimator of φ reduces to a matrix computations. To illustrate these points consider the case where $s = 0$ and consider the system $(\alpha I + \hat{T}^*\hat{T})\varphi = \hat{T}^*\hat{r}$, or equivalently:

$$\alpha\varphi + \frac{1}{n}\sum_{i=1}^{n} z_i\langle w_i, \frac{1}{n}\sum_{j=1}^{n} w_j\langle z_i, \varphi\rangle\rangle$$
$$= \frac{1}{n}\sum_{i=1}^{n} z_i\langle w_i, \frac{1}{n}\sum_{j=1}^{n} w_j y_j\rangle \tag{2.18}$$

This equation is estimated in two steps. First we take the scalar products of the two sides of this equation with any z_l ($l = 1, ..., n$) and we derive a linear system of n equations where the n unknowns are $\langle\varphi, z_l\rangle$. In a second step we may compute φ everywhere using the equation 2.17 and the previous computation of these scalar products. Note that we have to invert are $n \times n$ systems and that we assume to observe the scalar products $\langle z_i, z_j\rangle$ or $\langle w_i, w_j\rangle$ (and not necessarily the complete continuous trajectoires of the sample of W and Z).

The second question concerns the speed of convergence of the estimator. The main result is summarized by

$$\|\hat{\varphi}^\alpha - \varphi\|^2 \sim O(n^{-\frac{\beta}{\beta+1}}) \tag{2.19}$$

where $\beta = \frac{b}{a(1-\gamma)}$. We have defined b as the degree of smoothness of φ. The number a (the degree of ill-posedness) is defined by the property

$$\|T\varphi\| \sim \|L^{-a}\varphi\| \tag{2.20}$$

Intuitively L^{-1} is an integral operator and T is equivalent in terms of norms to L^{-a}.

The notation 2.19 is a shortcut of the property $C_1\|L^{-a}\varphi\| \leqslant \|T\varphi\| \leqslant C_2\|L^{-a}\varphi\|$ for two suitable constants C_1 and C_2.

The final term γ is specific to statistical inverse problems (different from inverse problems treated in numerical analysis). We have introduced an error term U and the element WU of $\mathscr{H}$ is assumed to have a variance Σ which is a trace class operator from $\mathscr{H}$ to $\mathscr{H}$. Let us consider the singular values decompositions of T^*T characterized by the (non zero) eigenvalues λ_j^2 and the eigenvectors φ_j. The parameter γ is defined by the largest value in $[0,1]$ such that

$$\sum_{j=1}^{\infty} \frac{\langle \Sigma \varphi_j, \varphi_j \rangle^2}{\lambda_j^{2\gamma}} < \infty \tag{2.21}$$

This property is trivially satisfied for $\gamma = 0$ because Σ is trace class which corresponds to the worst rate of convergence.

The last point we may consider concerns the optimality of our method in terms of the asymptotic variance. If we follow the result obtained in the finite dimensional case we should weight the difference $\hat{T}\varphi - \hat{r}$ in the norm by $\Sigma^{-\frac{1}{2}}$. This is impossible because Σ is a compact non invertible as a bounded operator. A second regularization could be used and we prove that the estimator

$$\hat{\varphi}^{\alpha,\nu} = \left(\alpha I + \hat{T}^* \hat{\Sigma}^{\frac{1}{2}} (\nu I + \hat{\Sigma})^{-2} \hat{\Sigma}^{\frac{1}{2}} \hat{T} \right)^{-1}$$
$$\hat{T}^* \hat{\Sigma}^{\frac{1}{2}} (\nu I + \hat{\Sigma})^{-2} \hat{\Sigma}^{\frac{1}{2}} \hat{r} \tag{2.22}$$

is optimal among a large class of estimators. All the elements of this class converge at the same rate and 2.21 has the best asymptotic variance. The study of this estimator is complex because it depends on two regularization parameters α and ν and because Σ is unknown and estimated. One of the basic results is that ν may be chosen such that the speed given in 2.18 is preserved.

2.3 The Additively Separable Model and Its Extensions

We still consider a random vector $(Y, Z, W) \in R \times R^p \times R^q$ and we define the instrumental regression by the following properties (see [13]):

$$\begin{cases} Y = \varphi(Z) + U \\ E(U|W) = 0 \end{cases} \tag{2.23}$$

if Z and W are identical 2.23 characterizes φ as the conditional expectation of Y given Z. The interest of the model comes from the case where Z and W are not identical and for simplicity we first assume that Z and W have no common element.

If (Y, Z, W) has a density f, the model 2.23 implies that φ should satisfy the integral equation:

$$\int \varphi(z) f(z|w) dz = \int y f(y|w) dy$$

which may be denoted

$$T\varphi = r \tag{2.24}$$

The choice of the spaces and then of the operator T and of r should be precised. In general we consider $L^2(Y, Z, W)$ Hilbert space of square integrable functions with respect to the true data generating process and $L^2(Z), L^2(W)$... the sub spaces of Z dependent or W dependent random variables. In that case r is assumed to be an element of $L^2(W)$, φ an element of $L^2(Z)$ and T is the conditional expectation operator from $L^2(Z)$ into $L^2(W)$. In that case the adjoint operator $T^*(L^2(W) \rightarrow L^2(Z))$ is simply the conditional expectation operator:

$$T^*(\psi) = E(\psi(W)|Z) \quad \psi \in L^2(W)$$

The difficulty behind this approach is that we have to estimate both r and T and we don't know the distribution which characterizes the spaces. It may be easier to specifies two given Hilbert spaces $\mathscr{E}$ and $\mathscr{F}$ and to assume that T operates from $\mathscr{E}$ to $\mathscr{F}$ and that $r \in \mathscr{F}$.

This approach has been follow in particular in [17]. In this presentation however we consider that the relevent spaces are of the form $L^2(Z)$....

The first question following from equation 2.24 is the identification of φ or equivalently the unicity of the solution. Due to the linearity of T it is obvious that φ is identified if $T\varphi = 0$ implies $\varphi = 0$.

This property is the injectivity of the conditional expectation operator and is a dependence condition between Z and W. It means that there does not exist a function of Z orthogonal to any function of W. This property has been introduced in statistics under the name "completeness" and has been studied under the name "strong identification" (see [19] chap. 5). For joint normal distribution T is one to one if and only if the rank of the covariance matrix between Z and W is equal to the dimension of Z.

In the general case the singular value decomposition of T may be used to characterize the identification condition. This condition is true if 0 is not an eigenvalue of T^*T.

Actually the statistical analysis of our problem requires that we characterize the speed of decline to zero of the SVD of T. This rate of decay is related to the dependence between Z and W. As we did in the previous section a natural tool is provided by a measurement deduced from an Hilbert scale defined from a differential operator L. We then assume that $\|T\varphi\| \sim \|L^{-a}\varphi\|$ and a defines the degree of ill-posedness of the problem. We may also assume that the singular values of T $(\lambda_j)_{j=1,...}$ decay at a geometric rate $(\lambda_j \sim \frac{1}{j^a})$ (see [24]) or at an exponential rate (which is the case for a jointly normal distribution from (Z, W)) .

As usual for non parametric statistic we need also to assume some regularity for the function we want to estimate. The Hilbert scale approach gives such a definition of regularity : φ has the regularity b if $\varphi \in \mathscr{D}(L^b)$. We can also assume some rate of decay for the Fourier coefficient of φ in the basis of the eigenvectors of T^*T. An elementary case is obtained by choosing $L = (T^*T)^{-\frac{1}{2}}$ which implies that the degree of ill-posedness is equal to 1 and the condition $\varphi \in \mathscr{D}[(T^*T)^{-\frac{b}{2}}]$ (or equivalently $\varphi \in \mathscr{R}(T^*T)^{\frac{b}{2}}$) is called a source condition. All these considerations are very common in the theory of inverse problems and we just applied this methodology to the conditional expectation operator.

The general principle of the estimation of φ is to estimate the r value of 2.24 and the operator T by usual non parametric technics and to solve any regularized version of equation 2.24.

The estimations is obtained by estimating the first order condition of the minimization of the Tikhonov functional (see [6]).

$$\|T\varphi - r\|^2 + \alpha\|\varphi\|^2 \tag{2.25}$$

which leads to

$$\varphi^\alpha = (\alpha I + T^*T)^{-1}T^*r \tag{2.26}$$

and to an estimator:

$$\hat{\varphi}^\alpha = (\alpha I + \hat{T}^*\hat{T})^{-1}\hat{T}^*\hat{r} \tag{2.27}$$

which may be computed by matrix inversion only (see [13]).

More general estimation are derived from iterated Tikhonov method or from a minimization in an Hilbert scale penalization:

$$\|T\varphi - r\|^2 + \alpha\|L^s\varphi\|^2 \tag{2.28}$$

where $s \leqslant \beta$, which leads to an estimator.

$$\hat{\varphi}^\alpha = L^{-s}(\alpha T + L^{-s}\hat{T}^*\hat{T}L^{-s})^{-1}L^{-1}\hat{T}^*\hat{r} \tag{2.29}$$

We will consider only the case where $s = 0$ (usual L^2 Tikhonov method).

Let us first discuss the non parametric estimation part. The rhs r may be estimated by a usual kernel approach:

$$\hat{r} = \frac{\sum_{i=1}^n y_i K\left(\frac{w-w_i}{h_n}\right)}{\sum_{i=1}^h K\left(\frac{w-w_i}{h_n}\right)} \tag{2.30}$$

where K is a kernel of suitable order an h_n the bandwidth. The estimation of T is done by replacing $f(z|w)$ by its kernel estimation

$$\hat{f}(z|w) = \frac{\frac{1}{h^p}\sum_{i=1}^n K\left(\frac{z-z_i}{h_n}\right)K\left(\frac{w-w_i}{h_n}\right)}{\sum_{i=1}^n K\left(\frac{w-w_i}{h_n}\right)} \tag{2.31}$$

where for simplicity we denote by K and h_n the different kernels and bandwidths. Then T is estimated by

$$\hat{T}\varphi = \int \varphi(z)\hat{f}(z|w)dz \tag{2.32}$$

and T^* by:

$$\hat{T}^*\varphi = \int \psi(w)\hat{f}(w|z)dw \tag{2.33}$$

where $\hat{f}(w|z)$ is defined analogously. Notice that $\hat{T}^*$ is not the dual of $\hat{T}$. It may be proved (see [9]) that:

$$\|\hat{r} - \hat{T}\varphi\|^2 \sim O\left(\frac{1}{nh_n^q} + h_n^{2\rho}\right) \tag{2.34}$$

$$\|\hat{T} - T\| \sim \|\hat{T}^* - T^*\|^2 \sim O\left(\frac{1}{nh_n^{p+q}} + h_n^{2\rho}\right) \tag{2.35}$$

where ρ represents the regularity of the joint distribution of the data.

Note that an alternative estimation of T would be

$$\hat{T}\varphi = \frac{\sum_{i=1}^{n}\varphi(z_i)K\left(\dfrac{w - w_i}{h_n}\right)}{\sum_{i=1}^{n}K\left(\dfrac{w - w_i}{h_n}\right)} \tag{2.36}$$

and equivalently for T^*. These estimations gives excellent approximations and excellent results in simulation but as this operators are not bounded in the L^2 spaces the available proofs of consistency do not apply to these estimations (see [12]).

As in the linear case, the computation of $\hat{\varphi}^\alpha$ reduces to matrix computation, at least if the approximation 2.36 is used. Indeed in that case we have to solve:

$$\alpha\varphi(z) + \frac{\sum_j \dfrac{\sum_i \varphi(z_i)K\left(\frac{w - w_i}{h_n}\right)}{\sum K\left(\frac{w_j - w_i}{h_n}\right)}K\left(\frac{z - z_j}{h_n}\right)}{\sum_j K\left(\frac{z - z_j}{h_n}\right)}$$

$$= \frac{\sum_j \dfrac{\sum_i y_i K\left(\frac{w_j - w_i}{h_n}\right)}{\sum K\left(\frac{w_j - w_i}{h_n}\right)}K\left(\frac{z - z_j}{h_n}\right)}{\sum_j K\left(\frac{z - z_j}{h_n}\right)} \tag{2.37}$$

which is solved in two steps: first for $z = z_1, ..., z_n$ and after for any value of z.

The last question is to consider the asymptotic properties of these estimators. Let us focus on the usual Tikhonov estimation. The difference $\hat{\varphi}^\alpha - \varphi$ may be decomposed in three parts:

$$
\begin{align}
\hat{\varphi}^* - \varphi &= (\alpha I + \hat{T}^*\hat{T})^{-1}\hat{T}^*(\hat{r} - \hat{T}\varphi) & \text{I} \\
&+ [(\alpha I + \hat{T}\hat{T})^{-1}\hat{T}^*\hat{T} - (\alpha I + T^*T)^{-1}T^*T]\varphi & \text{II} \\
&+ \varphi^\alpha - \varphi & \text{III}
\end{align}
$$

The norm $\|\varphi^\alpha - \varphi\|^2$ is the regularization bias and is known to be $O(\alpha^{\frac{b}{a}})$ in the Hilbert scale approach.

The norm of the first term I verifies

$$
\|\text{I}\|^2 \leqslant \|(\alpha I + \hat{T}^*\hat{T})^{-1}\hat{T}^*\|^2\|\hat{r} - \hat{T}\varphi\|^2
$$

$$
\sim O\left(\frac{1}{\alpha}\left(\frac{1}{nh^p} + h^{2\rho}\right)\right)
$$

The norm of II requires some computations but under some regularity assumption is term is negligible with respect to to the other term.

If h is chosen by an optimal rule we have $h_n = n^{-\frac{1}{p+2\rho}}$ and $\|\text{I}\|^2 \sim O\left(\frac{1}{\alpha}n^{-\frac{2\rho}{p+2\rho}}\right)$.

The optimal choice for α is then

$$
\alpha \text{ proportional to } n^{-\left[\frac{2\rho}{p+2\rho}\frac{a}{b+a}\right]} \tag{2.38}
$$

which gives an optimal rate of convergence:

$$
\|\hat{\varphi}^\alpha - \varphi\|^2 \sim O(n^{-\left[\frac{2\rho}{p+2\rho} \times \frac{b}{a+b}\right]})
$$

In some cases (see [7], [16], [30]) it is natural to assume that $\rho = b + a$ and the optimal rate simplifies to:

$$
\|\hat{\varphi}^\alpha\|^2 \sim O(n^{-\frac{2b}{2(b+a)+p}}) \tag{2.39}
$$

which has been shown to be minimax under some assumptions.

The main question following from this approach is the empirical determination of the regularization parameters, namely the bandwidths of the kernel estimation and the α for the Tikhonov regularization.

Several approaches have been proposed in the literature. The following rule has been proved to have good properties, both theoretically and by simulation (see [10] or [12]).

The principle is to compute α which minimizes

$$
\frac{1}{\alpha}\|\hat{r} - \hat{T}\hat{\varphi}^\alpha\|^2 \tag{2.40}
$$

where the norm is replaced by the empirical norm.

Indeed the minimization of $\|\hat{r} - \hat{T}\hat{\varphi}^{\alpha}\|^2$ leads to $\alpha = 0$ and multiplying by $\frac{1}{\alpha}$ is equivalent to penalize this quantity. The α obtained by this rule has the optimal rate of convergence (for a comparable rule see [33] or for a Bayesian approach in [14]) and numerous simulations show its relevance.

This separable model has many extensions which may be treated in the same spirit.

First, for dimensionality reason, we may consider some restrictions on the general form $Y = \varphi(Z) + U$, for example:

i) $Y = \varphi_1(Z_1) + \varphi_2(Z_2) + U$ (additive model) $Y = \varphi_1(Z_1) + Z_2'\beta_2 + U$ (semi para-metric additive model) where Z_2 may be exogenous (Z_2 included in W) or not

ii) $Y = \varphi(\beta'Z) + U$ (single index form). These models do not lead to a linear integral equation and have been treated in many papers (see [17], [1]).

Secondly we may consider the class of transformation models like:

$$\varphi(Y) = Z'\beta + U$$

(see [12]) or

$$\varphi(Y) = \psi(Z) + X'\beta + U$$

(see [21] when X is exogenous).

Third we may consider some test problems like testing that Z is exogenous ($\varphi(Z) = E(Y|Z)$) or that φ has a given parametric form (see [28], [3]).

2.4 The Non-separable Models

The last family of models we consider in the static case belong to the class of non separable models. Let us still consider a random vector $(Y, Z, W) \in R \times R^p \times R^q$ and we assume the following relation:

$$
\begin{aligned}
&Y = \varphi(Z, U) \quad U \in R \\
&\text{where } \varphi(Z, .) \text{ is strictly increasing.} \\
&U \perp\!\!\!\perp W \text{ and } U \sim F_0 \text{ given}
\end{aligned}
\tag{2.41}
$$

If $Z = W$ and U uniform this model is called the conditional quantile model and may be viewed as a way to describe the conditional distribution of Y given Z. If U is exponential and Y non negative this equation is a general characterization of duration model conditional on cofactors Z (see [27]). This model may be generalized by relaxing some assumptions as the monotonicity condition.

Our objective here is to relax the assumption $Z = W$ and to consider the instrumental variable generalization of the non separable models by considering the case where Z and W are distincts (see [29]). A complete theory of these models is out of the scope of this survey but we want to give some elements on this specification. A good example of such a model comes from duration models. Remember that if Y is a duration with a one to one integrated hazard rate Λ we have the relation $Y = \Lambda^{-1}(U) = \varphi(U)$ where U is an exponential distribution with parameter egal to 1. The extension of this model to exogenous explanatory variables or cofactors considers an equation $Y = \varphi(Z, U)$ where Z and U are independent. In numerous models this assumption is not realistic (in particular if Z is a treatment assigned non independently of U) and may be replaced by the independence between U and some instruments W.

First let us note that equation 2.41 leads on non linear integral equation where the unknown element is the fonction φ. The methodology we follow for analysis this problem is to start with the condition $U \perp\!\!\!\perp W$ equivalent to $U|W \sim F_0$ and we easily show that this condition may be written as a property of the joint cumulative distribution of Y and density of Z conditionally on W.

$$U \perp\!\!\!\perp W \Leftrightarrow \int Prob(U \leqslant u, Z = z | W = w)dz = F_0(u)$$

$$\Leftrightarrow \int F(\varphi(z,u), z|w)dz = F_0(u) \tag{2.42}$$

where $F(y, z|w) = Prob(Y \leqslant y, Z = z | W = w)$

$$= \frac{\partial^p}{\partial z_1, ..., \partial z_p} Prob(Y \leqslant y, Z \leqslant z | W = w) \tag{2.43}$$

The fonction F is identifiable and estimable from the data, F_0 the c.d.f. of U is given and 2.42 appears as an equation which characterizes φ.

The next question is the question of identification, i.e. the unicity of the solution of 2.42. As the equation is non linear it is natural to look at the local unicity of the solution which may be characterized by the one to one property of the linear approximation of the equation at the true value.

Let $f(y, z|w)$ the density of (Y, Z) given $W = w$ $(f(y, z|w) = \frac{\partial}{\partial y} F(y, z|w))$. The linearized version of the equation 2.42 denoted $T(\varphi) = F_0$ is based on the linear operator :

$$T'_\varphi(\tilde{\varphi}) = \int \tilde{\varphi}(z, u) f(\varphi(z, u), z|w)dz \tag{2.44}$$

This operator is computed as the Gâteau derivative of T in φ and is shown to be the Frechet derivative under regularity conditions (see [34]). In particular T' may be assumed continuous if its image space is provided with a suitable topology. The model is then locally identified if T'_φ is one to one for any φ. Assuming that the true φ is almost surely (as a function of Z and U) different from 0 we have:

$$T'_\varphi(\tilde{\varphi}) = 0 \Leftrightarrow \int \frac{\tilde{\varphi}(z, u)}{\varphi(z, u)} \varphi(z, u) f(\varphi(z, u), z|w)dz = 0$$

$$\Leftrightarrow g(u|w) \int \frac{\tilde{\varphi}(z, u)}{\varphi(z, u)} g(z|w, u)dz = 0 \tag{2.45}$$

if $g(z, u|w)$ is the density of (Z, U) given W.

We say that Z is strongly identified by W given U if for any integrable function $\lambda(Z,U)$ we have $E(\lambda(Z,U)|W,U) = 0$ implies $\lambda(Z,U) = 0$ almost surely. It follows immediately that φ is locally identified if Z is strongly identified by W given U.

Let us now briefly discuss the estimation procedure of φ. The principle would be to construct a regularized solution of

$$min\|T(\varphi) - F_0\|^2 \qquad (2.46)$$

where T is replaced by a non parametric estimator. This minimization is difficult and may lead to unconsistent estimator for some estimators of T. A better strategy is to estimate the first order conditions of the Tikhonov functional

$$\|T(\varphi) - F_0\|^2 + \alpha\|\varphi\|^2 \qquad (2.47)$$

i.e.

$$\alpha\varphi + T_\varphi^{'*}(T(\varphi) - F_0) = 0 \qquad (2.48)$$

where $T_\varphi^{'*}$ is the adjoint of $T_\varphi^{'}$ defined in 2.41. We have:

$$T_\varphi^{'*}(\psi) = \int \psi(w)f(\varphi(z,u),w|z)dw$$

where $f(y,w|z)$ is the joint density of (Y,W) given Z.

Numerous iterative methods exists for solving a non linear integral equation and are out of the scope of this chapter (see [31] Kaltenbacher et al (2008) for a survey of these methods).

For example we may consider the following iterative method. If $\hat{\phi}_{k-1}^\alpha$ is the value of the estimator at step $k - 1$ the new value $\hat{\phi}_k^\alpha$ will be the solution of

$$\alpha\varphi + T_{\hat{\phi}_{k-1}^\alpha}^{'\alpha}(T(\varphi) - F_0)) = 0 \qquad (2.49)$$

The parameter α may be fixed or updated at each step. The algorithm is stopped at the convergence ($\hat{\phi}_{k-1}^\alpha \simeq \hat{\phi}_k^\alpha$) because the regularization is coming from the α parameter.

We don't consider in that section the extension of the analysis of the convergence rate of the estimator of φ neither then the optimal selection of the regularization parameter (see [23], [8] or [29]).

2.5 Some Extensions to Dynamic Models

All the specifications we have considered have been introduced in an $i.i.d.$ context. Their extension to some dynamic case with discrete time observations is natural. Take for example the model

$$Y_t = \varphi(Z_t) + U_t \qquad (2.50)$$

where (Y_t, Z_t) is a markov process and

$$E(U|Y_{t-1}, Z_{t-1}) = 0 \tag{2.51}$$

All the theory of section 2.3 applies with instruments W that are lagged variables (Y_{t-1}, Z_{t-1}). In case of weakly dependent processes the main results of non parametric estimation apply and for example the analysis of the rate of convergence remains identical. Non stationary data mainly leads to unexpected conclusions: if (Y_t, Z_t) is a unit root process, [37] verify that a usual estimation of the regression of Y_t given Z_t is a consistent estimators of φ.

We want to give a brief survey of a more theoretical approach for instrumental analysis for stochastic processes, possibly with a continuous time. A complete presentation and examples are given in [20].

We will briefly present two different approaches which correspond for stochastic processes to the extension of separable and non separable models.

We reproduce here the general approach we have followed through this chapter. A process $(Y_t)_t$ is written as a function of two processes $(Z_t)_t$ and (U_t). The $(Z_t)_t$ process is observable and represents the endogenous variables and $(U_t)_t$ is an observable dynamic noise. We consider a third process $(W_t)_t$ of instruments. We will write Y function of Z and U in two possible forms and we complete this specification by "independence-type" conditions between U and W. The first one will be that U_t verifies a martingale condition with respect of W_t: the increment of $U_t - U_s$ are assumed to have a zero conditional mean given the past of W until s. The second one is a complete independence between the two processes U and W. Many possible relations between Y and (Z, U) may be constructed: the two particular one are specially relevant for usual models.

The first extension considers a stochastic process Y_t $(t \geqslant 0)$ possibly with t continuous and two filtrations $\mathscr{Z}_t$ and $\mathscr{W}_t$. In general there exists two stochastic processes Z_t and W_t such that $\mathscr{Z}_t$ is generated by Y_t and Z_t and $\mathscr{W}_t$ by Y_t and W_t. In an intuitive presentation the idea is to decompose the variation of Y_t in this way:

$$dY_t = \lambda_t dt + dU_t \tag{2.52}$$

where λ_t depends on $\mathscr{Z}_t$ and where $E(dU_t|\mathscr{W}_t) = 0$. More formally if we integrate with respect to to t equation 2.52 we get

$$Y_t = \Lambda_t + U_t \tag{2.53}$$

where Λ_t is $\mathscr{Z}_t$ predictable and U_t satisfies the martingale condition $E(U_t - U_s|\mathscr{W}_s) = 0$.

This model may be identified by computing first the decomposition of Y_t with respect to to $\mathscr{W}_t$:

$$Y_t = H_t + M_t \tag{2.54}$$

where H_t is $\mathscr{W}_t$ predictable and M_t is a $\mathscr{W}_t$ martingale. We assume that 2.54 has a differential version

$$dY_t = h_t dt + dM_t \tag{2.55}$$

In that case λ_t is solution of

$$h_t = E(\lambda_t | \mathscr{W}_t) \quad \forall t. \tag{2.56}$$

The equation 2.56 generates a sequence of linear integral equations which may be treated (for each value of t) in the same way as in section 2.3. However h_t and λ_t may depend on the complete past of Y_t and W_t for h_t and of Y_t and Z_t for λ_t and the statistical treatment of this problem is impossible unless some restrictions are imposed to these processes.

This decomposition does not cover all the interesting cases and we propose another class of stochastic processes models with endogenous variables defined in the following way.

Let φ_t an increasing sequence of stopping times adapted to the filtration $\mathscr{Z}_t$ and U_t a process with a given distribution. We assume $(U_t)_t$ and $(W_t)_t$ independent (the complete paths of U and W are independent) and the model is defined by assuming

$$Y_{\varphi_t} = U_t, \tag{2.57}$$

i.e. the process Y stopped at φ_t is equal to U_t.

This model may be viewed as a non separable model and be used for counting processes (U_t is an homogenous Poisson process) or for diffusion (U_t is a Brownian motion). It is shown in [20] that φ_t is characterized as the solution of a non linear integral equation:

$$\int Q(dz) \int_0^{\varphi_t} k_t g(z | \mathscr{W}_s) ds = H_t^U \tag{2.58}$$

where H_t^U is the compensator U_t with respect to its own his history and is given, k_t is the intensity of Y_t with respect to $\mathscr{Z}_t$ and $\mathscr{W}_t$ and g is the density of the process $(Z_t)_t$ with respect to a dominating measure Q.

Several examples of the application of this formulae are given in [20]. In this paper the local unicity of the solution of this sequence of equations is also discussed.

2.6 Conclusion

This chapter present different examples of econometric models based on an instrument variable assumption and shows that the functional parameter of interest is characterized as the solution of a linear or non linear integral equation. We have illustrated the main questions following of this characterization: unicity or local unicity of the solution, degree of ill-posedness and regularization, rate of convergence of the solutions and data driven selection of the regularization parameters. All these points have been illustrated by a Monte Carlo analysis in [12] and an application may be founded e.g. in [2].

References

1. Ai, C., Chen, X.: Efficient estimation of models with conditional moment restrictions containing unknown functions. Econometrica **71**(6), 1795–1843 (2003)
2. Blundell, R., Chen, X., Kristensen, D.: Semi-nonparametric IV estimation of shape-invariant Engel curves. Econometrica **75**(6), 1613–1669 (2007)
3. Blundell, R., Horowitz, J.: A non-parametric test of exogeneity. Rev. Econom. Stud. **74**(4), 1035–1058 (2007)
4. Carrasco, M.: A regularization approach to the many instruments problem. Tech. rep., Mimeo University of Montreal (2008)
5. Carrasco, M., Florens, J.P.: Generalization of the GMM in presence of a continuum of moment conditions. Econometric Theory **16**, 797–834 (2000)
6. Carrasco, M., Florens, J.P., Renault, E.: Linear inverse problems in structural econometrics: estimation based on spectral decomposition and regularization. In: J. Heckman, E. Leamer (eds.) Handbook of Econometrics, vol. 6B. North Holland, Amsterdam (2007)
7. Chen, X., Reiss, M.: On rate optimality for ill-posed inverse problems in econometrics. Econ. Theory (2010). In press
8. Chernozhukov, V., Gagliardini, P., Scaillet, O.: Nonparametric instrumental variable estimators of quantile structural effects. Tech. rep., HEC, Swiss finance institute (2009)
9. Darolles, S., Florens, J.P., Renault, E.: Non parametric instrumental regression. Tech. rep., IDEI working paper. Econometrica (forthcoming) (2003)
10. Engl, H., Hanke, M., Neubauer, A.: Regularization of Inverse Problems. Kluwer, Dordrecht (2000)
11. Engle, R., Hendry, D., Richard, J.F.: Exogeneity. Econometrica **51**(2), 277–304 (1983)
12. Fève, F., Florens, J.P.: The practice of nonparametric estimation by solving inverse problem: the example of transformation models. Econom. J. **13**(3), S1–S27 (2010)
13. Florens, J.P.: Inverse problems and structural econometrics: The example of instrumental variables. In: L.H. M. Dewatripont, S. Turnovsky (eds.) Advances in Economics and Econometrics: Theory and Applications, pp. 284–311. Cambridge University Press (2003)
14. Florens, J.P., Simoni, A.: Regularized posteriors in linear ill-posed inverse problems (2010). Discussion paper
15. Florens, J.P., Heckman, J., Meghir, C., Vytlacil, E.: Identification of treatment effects using control functions in models with continuous, endogenous treatment and heterogeneous effects. Econometrica **76**(5), 1191–1206 (2008)
16. Florens, J.P., Johannes, J., Van Bellegem, S.: Identification and estimation by penalization in nonparametric instrumental regression. Econ. Theory (2010). In press
17. Florens, J.P., Johannes, J., Van Bellegem, S.: Instrumental regression in partially linear models. Econometrics J. **13**, S1–S27 (2010)
18. Florens, J.P., Mouchart, M.: Conditioning in dynamic models. J. Time Series Anal. **53**(1), 15–35 (1985)
19. Florens, J.P., Mouchart, M., Rolin, J.: Elements of Bayesian Statistics. Dekker, New York (1990)
20. Florens, J.P., Simon, G.: Endogeneity and instrumental variables in dynamic models (2010). Working paper TSE
21. Florens, J.P., Sokullu, S.: Semiparametric transformation models (2010). Working paper TSE
22. Florens, J.P., Van Bellegem, S.: Functional instrumental linear regression (2010). Mimeo, University of Toulouse
23. Gagliardini, C., Scaillet, O.: Tikhonov regularisation for functional minimum distance estimators (2006). Discussion paper
24. Hall, P., Horowitz, J.: Nonparametric methods for inference in the presence of instrumental variables. Ann. Statist. **33**, 2904–2929 (2005)
25. Hall, P., Horowitz, J.: Methodology and convergence rates for functional linear regression. Ann. Statist. **35**, 70–91 (2007)

26. Hansen, L.: Large sample properties of generalized methods of moments estimators. Econometrica **50**, 1029–1054 (1982)
27. Horowitz, J.: Semiparametric estimation of a regression model with an unknown transformation of the dependent variable. Econometrica **64**, 103–137 (1996)
28. Horowitz, J.: Testing a parametric model against a nonparametric alternative with identification through instrumental variables. Econometrica **74**, 521–538 (2006)
29. Horowitz, J., Lee, S.: Non parametric instrumental variables estimation of a quantile regression model. Econometrica **75**, 1191–1208 (2007)
30. Johannes, J., Van Bellegem, S., Vanhems, A.: A unified approach to solve ill-posed inverse problems in econometrics (2010). In press
31. Kaltenbacher, B., Neubauer, A., Scherzer, O.: Iterative regularization methods for non linear ill-posed problems. De Grugter, Berlin (2008)
32. Koopmans, T., Reiersol, O.: The identification of structural characteristics. Ann. Math. Statist. **21**, 165–181 (1950)
33. Loubes, J., Marteau, C.: Paths toward adaptive estimation for instrumental variables regression (2010). Working paper IMT
34. Nashed, M.: Generalized inverses, normal solvability and iterations for singular operators equations. In: L. Rall (ed.) Nonlinear functional analysis and application, pp. 311–359. Academic Press (1971)
35. Newey, W., Powell, J., Vella, F.: Nonparametric estimation of triangular simultaneous equations models. Econometrica **67**, 565–604 (1999)
36. Stock, J., Wright, J., Yogo, M.: A survey of weak instruments and weak identification in generalized method of moments. J. Bus. Econom. Statist. **20**(4), 518–529 (2002)
37. Wang, Q., Phillips, P.: Structural non parametric cointegration regression. Econometrica **77**(6), 1901–1948 (2009)

Part III
Lecture Notes on High-Dimensional Estimation

Chapter 3
High Dimensional Sparse Econometric Models: An Introduction

Alexandre Belloni and Victor Chernozhukov

Abstract In this chapter we discuss conceptually high dimensional sparse econometric models as well as estimation of these models using ℓ_1-penalization and post-ℓ_1-penalization methods. Focusing on linear and nonparametric regression frameworks, we discuss various econometric examples, present basic theoretical results, and illustrate the concepts and methods with Monte Carlo simulations and an empirical application. In the application, we examine and confirm the empirical validity of the Solow-Swan model for international economic growth.

3.1 The High Dimensional Sparse Econometric Model

We consider linear, high dimensional sparse (HDS) regression models in econometrics. The HDS regression model has a large number of regressors p, possibly much larger than the sample size n, but only a relatively small number $s < n$ of these regressors are important for capturing accurately the main features of the regression function. The latter assumption makes it possible to estimate these models effectively by searching for approximately the right set of the regressors, using ℓ_1-based penalization methods. In this chapter we will review the basic theoretical properties of these procedures, established in the works of [8, 10, 18, 17, 7, 15, 13, 26, 25], among others (see [20, 7] for a detailed literature review). In this section, we review the modeling foundations as well as motivating examples for these procedures, with emphasis on applications in econometrics.

Let us first consider an exact or parametric HDS regression model, namely,

Alexandre Belloni
Duke University, Fuqua School of Business, 100 Fuqua Drive, Durham, NC 27708, USA, e-mail: abn5@duke.edu

Victor Chernozhukov
Massachusetts Institute of Technology, Department of Economics, 50 Memorial Drive, Cambridge, MA 02142, USA, e-mail: vchern@mit.edu

$$y_i = x_i'\beta_0 + \varepsilon_i, \quad \varepsilon_i \sim N(0, \sigma^2), \quad \beta_0 \in \mathbb{R}^p, \quad i = 1, \ldots, n, \qquad (3.1)$$

where y_i's are observations of the response variable, x_i's are observations of p-dimensional fixed regressors, and ε_i's are i.i.d. normal disturbances, where possibly $p \geqslant n$. The key assumption of the exact model is that the true parameter value β_0 is sparse, having only $s < n$ non-zero components with support denoted by

$$T = \text{support}(\beta_0) \subset \{1, \ldots, p\}. \qquad (3.2)$$

Next let us consider an approximate or nonparametric HDS model. To this end, let us introduce the regression model

$$y_i = f(z_i) + \varepsilon_i, \quad \varepsilon_i \sim N(0, \sigma^2), \quad i = 1, \ldots, n,$$

where y_i is the outcome, z_i is a vector of elementary fixed regressors, $z \mapsto f(z)$ is the true, possibly non-linear, regression function, and ε_i's are i.i.d. normal disturbances. We can convert this model into an approximate HDS model by writing

$$y_i = x_i'\beta_0 + r_i + \varepsilon_i, \quad i = 1, \ldots, n,$$

where $x_i = P(z_i)$ is a p-dimensional regressor formed from the elementary regressors by applying, for example, polynomial or spline transformations, β is a conformable parameter vector, whose "true" value β_0 has only $s < n$ non-zero components with support denoted as in (3.2), and $r_i := r(z_i) = f(z_i) - x_i'\beta_0$ is the approximation error. We shall define the true value β_0 more precisely in the next section. For now, it is important to note only that we assume there exists a value β_0 having only s non-zero components that sets the approximation error r_i to be small.

Before considering estimation, a natural question is whether exact or approximate HDS models make sense in econometric applications. In order to answer this question it is helpful to consider the following example, in which we abstract from estimation completely and only ask whether it is possible to accurately describe some structural econometric function $f(z)$ using a low-dimensional approximation of the form $P(z)'\beta_0$. In particular, we are interested in improving upon the conventional low-dimensional approximations.

Example 1: Sparse Models for Earning Regressions. In this example we consider a model for the conditional expectation of log-wage y_i given education z_i, measured in years of schooling. Since measured education takes on a finite number of years, we can expand the conditional expectation of wage y_i given education z_i:

$$E[y_i|z_i] = \sum_{j=1}^{p} \beta_{0j} P_j(z_i), \qquad (3.3)$$

using some dictionary of approximating functions $P_1(z_i), \ldots, P_p(z_i)$, such as polynomial or spline transformations in z_i and/or indicator variables for levels of z_i. In fact, since we can consider an overcomplete dictionary, the representation of the function may not be unique, but this is not important for our purposes.

A conventional sparse approximation employed in econometrics is, for example,

$$f(z_i) := E[y_i|z_i] = \tilde{\beta}_1 P_1(z_i) + \cdots + \tilde{\beta}_s P_s(z_i) + \tilde{r}_i, \tag{3.4}$$

where the P_j's are low-order polynomials or splines, with typically $s = 4$ or 5 terms, but there is no guarantee that the approximation error $\tilde{r}_i$ in this case is small, or that these particular polynomials form the best possible s-dimensional approximation. Indeed, we might expect the function $E[y_i|z_i]$ to exhibit oscillatory behavior near the schooling levels associated with advanced degrees, such as MBA or MD. Low-degree polynomials may not be able to capture this behavior very well, resulting in large approximation errors $\tilde{r}_i$'s.

Therefore, the question is: With the same number of parameters, can we find a much better approximation? In other words, can we find some higher-order terms in the expansion (3.3) which will provide a higher-quality approximation? More specifically, can we construct an approximation

$$f(z_i) := E[y_i|z_i] = \beta_{k_1} P_{k_1}(z_i) + \cdots + \beta_{k_s} P_{k_s}(z_i) + r_i, \tag{3.5}$$

for some regressor indices $k_1, \ldots, k_s$ selected from $\{1, \ldots, p\}$, that is accurate and much better than (3.4), in the sense of having a much smaller approximation error r_i?

Obviously the answer to the latter question depends on how complex the behavior of the true regression function (3.3) is. If the behavior is not complex, then low-dimensional approximation should be accurate. Moreover, it is clear that the second approximation (3.5) is weakly better than the first (3.4), and can be much better if there are some important high-order terms in (3.3) that are completely missed by the first approximation. Indeed, in the context of the earning function example, such important high-order terms could capture abrupt positive changes in earning associated with advanced degrees such as MBA or MD. Thus, the answer to the question depends strongly on the empirical context.

Consider for example the earnings of prime age white males in the 2000 U.S. Census (see e.g., Angrist, Chernozhukov and Fernandez-Val [2]). Treating this data as the population data, we can then compute $f(z_i) = E[y_i|z_i]$ without error. Figure 3.1 plots this function. (Of course, such a strategy is not generally available in the empirical work, since the population data are generally not available.) We then construct two sparse approximations and also plot them in Figure 3.1: the first is the conventional one, of the form (3.4), with $P_1, \ldots, P_s$ representing an $(s-1)$-degree polynomial, and the second is an approximation of the form (3.5), with $P_{k_1}, \ldots, P_{k_s}$ consisting of a constant, a linear term, and two linear splines terms with knots located at 16 and 19 years of schooling (in the case of $s = 5$ a third knot is located at 17). In fact, we find the latter approximation automatically using ℓ_1-penalization methods, although in this special case we could construct such an approximation just by eye-balling Figure 3.1 and noting that most of the function is described by a linear function, with a few abrupt changes that can be captured by linear spline terms that induce large changes in slope near 17 and 19 years of schooling. Note that an exhaustive search for a low-dimensional approximation requires looking at a very large set of models. We avoided this exhaustive search by using ℓ_1-penalized least squares (LASSO), which penalizes the size of the model through the sum of absolute values of regression coefficients. Table 3.1 quantifies the performance of the different sparse approximations. (Of course, a simple strategy of eye-balling also works in this simple illustrative setting, but clearly does not apply to more general examples with several conditioning variables z_i, for example, when we want to condition on education, experience, and age.)

Table 3.1 Errors of Conventional and the LASSO-based Sparse Approximations of the Earning Function. The LASSO estimator minimizes the least squares criterion plus the ℓ_1-norm of the coefficients scaled by a penalty parameter λ. As shown later, it turns out to have only a few non-zero components. The Post-LASSO estimator minimizes the least squares criterion over the non-zero components selected by the LASSO estimator.

Sparse Approximation	s	L_2 error	L_∞ error
Conventional	4	0.1212	0.2969
Conventional	5	0.1210	0.2896
LASSO	4	0.0865	0.1443
LASSO	5	0.0752	0.1154
Post-LASSO	4	0.0586	0.1334
Post-LASSO	5	0.0397	0.0788

The next two applications are natural examples with large sets of regressors among which we need to select some smaller sets to be used in further estimation and inference. These examples illustrate the potential wide applicability of HDS modeling in econometrics, since many classical and new data sets have naturally multi-dimensional regressors. For example, the American Housing Survey records prices and multi-dimensional features of houses sold, and scanner data-sets record prices and multi-dimensional information on products sold at a store or on the internet.

Example 2: Instrument Selection in Angrist and Krueger Data. The second example we consider is an instrumental variables model, as in Angrist and Krueger [3]

$$y_{i1} = \theta_0 + \theta_1 y_{i2} + w_i'\gamma + v_i, \; E[v_i|w_i,x_i] = 0,$$
$$y_{i2} = x_i'\beta + w_i'\delta + \varepsilon_i, \qquad E[\varepsilon_i|w_i,x_i] = 0,$$

where for person i, y_{i1} denotes wage, y_{i2} denotes education, w_i denotes a vector of control variables, and x_i denotes a vector of instrumental variables that affect education but do not directly affect the wage. The instruments x_i come from the quarter-of-birth dummies, and from a very large list, total of 180, formed by interacting quarter-of-birth dummies with control variables w_i. The interest focuses on measuring the coefficient θ_1, which summarizes the causal impact of education on earnings, via instrumental variable estimators.

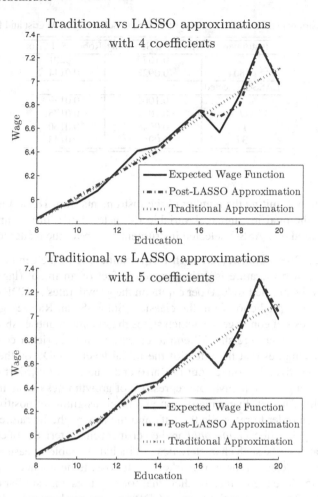

Fig. 3.1 The figures illustrates the Post-LASSO sparse approximation and the traditional (low degree polynomial) approximation of the wage function. The top figure uses $s = 4$ and the bottom figure uses $s = 5$.

There are two basic options used in the literature: one uses just the quarter-of-birth dummies, that is, the leading 3 instruments, and another uses all 183 instruments. It is well known that using just 3 instruments results in estimates of the schooling coefficient θ_1 that have a large variance and small bias, while using 183 instruments results in estimates that have a much smaller variance but (potentially) large bias, see, e.g., [14]. It turns out that, under some conditions, by using ℓ_1-based estimation of the first stage, we can construct estimators that also have a nearly efficient variance and at the same time small bias. Indeed, as shown in Table 3.2, using the LASSO estimator induced by different penalty levels defined in Section 3.2, it is possible to find just 37 instruments that contain nearly all information in the first

Table 3.2 Instrumental Variable Estimates of Return to Schooling in Angrist and Krueger Data

Instruments	Return to Schooling	Robust Std Error
3	0.1077	0.0201
180	0.0928	0.0144
LASSO-selected		
5	0.1062	0.0179
7	0.1034	0.0175
17	0.0946	0.0160
37	0.0963	0.0143

stage equation. Limiting the number of the instruments from 183 to just 37 reduces the bias of the final instrumental variable estimator. For a further analysis of IV estimates based on LASSO-selected instruments, we refer the reader to [6].

Example 3: Cross-Country Growth Regression. One of the central issues in the empirical growth literature is estimating the effect of an initial (lagged) level of GDP (Gross Domestic Product) per capita on the growth rates of GDP per capita. In particular, a key prediction from the classical Solow-Swan-Ramsey growth model is the hypothesis of convergence, which states that poorer countries should typically grow faster and therefore should tend to catch up with the richer countries. Such a hypothesis implies that the effect of the initial level of GDP on the growth rate should be negative. As pointed out in Barro and Sala-i-Martin [5], this hypothesis is rejected using a simple bivariate regression of growth rates on the initial level of GDP. (In this data set, linear regression yields an insignificant positive coefficient of 0.0013.) In order to reconcile the data and the theory, the literature has focused on estimating the effect *conditional* on the pertinent characteristics of countries. Covariates that describe such characteristics can include variables measuring education and science policies, strength of market institutions, trade openness, savings rates and others [5]. The theory then predicts that for countries with similar other characteristics the effect of the initial level of GDP on the growth rate should be negative ([5]). Thus, we are interested in a specification of the form:

$$y_i = \alpha_0 + \alpha_1 \log G_i + \sum_{j=1}^{p} \beta_j X_{ij} + \varepsilon_i, \qquad (3.6)$$

where y_i is the growth rate of GDP over a specified decade in country i, G_i is the initial level of GDP at the beginning of the specified period, and the X_{ij}'s form a long list of country i's characteristics at the beginning of the specified period. We are interested in testing the hypothesis of convergence, namely that $\alpha_1 < 0$.

Given that in standard data-sets, such as Barro and Lee data [4], the number of covariates p we can condition on is large, at least relative to the sample size n, covariate selection becomes a crucial issue in this analysis ([16], [21]). In particular, previous findings came under severe criticism for relying on ad hoc procedures for covariate selection. In fact, in some cases, all of the previous findings have been questioned ([16]). Since the number of covariates is high, there is no simple way to resolve the model selection problem using only classical tools. Indeed the number of possible lower-dimensional models is very large, although [16] and [21] attempt to search over several millions of these models. We suggest ℓ_1-penalization and post-ℓ_1-penalization methods to address this important issue. In Section 3.8, using these methods we estimate the growth model (3.6) and indeed find rather strong support for the hypothesis of convergence, thus confirming the basic implication of the Solow-Swan model.

Notation. In what follows, all parameter values are indexed by the sample size n, but we omit the index whenever this does not cause confusion. In making asymptotic statements, we assume that $n \to \infty$ and $p = p_n \to \infty$, and we also allow for $s = s_n \to \infty$. We use the notation $(a)_+ = \max\{a, 0\}$, $a \vee b = \max\{a, b\}$ and $a \wedge b = \min\{a, b\}$. The ℓ_2-norm is denoted by $\| \cdot \|$ and the "ℓ_0-norm" $\| \cdot \|_0$ denotes the number of non-zero components of a vector. Given a vector $\delta \in \mathbb{R}^p$, and a set of indices $T \subset \{1, \ldots, p\}$, we denote by δ_T the vector in which $\delta_{Tj} = \delta_j$ if $j \in T$, $\delta_{Tj} = 0$ if $j \notin T$. We also use standard notation in the empirical process literature,

$$\mathbb{E}_n[f] = \mathbb{E}_n[f(w_i)] = \sum_{i=1}^{n} f(w_i)/n,$$

and we use the notation $a \lesssim b$ to denote $a \leqslant cb$ for some constant $c > 0$ that does not depend on n; and $a \lesssim_P b$ to denote $a = O_P(b)$. Moreover, for two random variables X, Y we say that $X =_d Y$ if they have the same probability distribution. We also define the prediction norm associated with the empirical Gram matrix $\mathbb{E}_n[x_i x_i']$ as

$$\|\delta\|_{2,n} = \sqrt{\mathbb{E}_n[(x_i'\delta)^2]}.$$

3.2 The Setting and Estimators

3.2.1 The Model

Throughout the rest of the chapter we consider the nonparametric model introduced in the previous section:

$$y_i = f(z_i) + \varepsilon_i, \quad \varepsilon_i \sim N(0, \sigma^2), \quad i = 1, \ldots, n, \tag{3.7}$$

where y_i is the outcome, z_i is a vector of fixed regressors, and ε_i's are i.i.d. disturbances. Define $x_i = P(z_i)$, where $P(z_i)$ is a p-vector of transformations of z_i, including a constant, and $f_i = f(z_i)$. For a conformable sparse vector β_0 to be defined below, we can rewrite (3.7) in an approximately parametric form:

$$y_i = x_i'\beta_0 + u_i, \quad u_i = r_i + \varepsilon_i, \quad i = 1, \ldots, n, \tag{3.8}$$

where $r_i := f_i - x_i'\beta_0$, $i = 1, \ldots, n$, are approximation errors. We note that in the parametric case, we may naturally choose $x_i'\beta_0 = f_i$ so that $r_i = 0$ for all $i = 1, \ldots, n$. In the nonparametric case, we shall choose $x_i'\beta_0$ as a sparse parametric model that yields a good approximation to the true regression function f_i in equation (3.7).

Given (3.8), our target in estimation will become the parametric function $x_i'\beta_0$. Here we emphasize that the ultimate target in estimation is, of course, f_i, while $x_i'\beta_0$ is a convenient intermediate target, introduced so that we can approach the estimation problem as if it were parametric. Indeed, the two targets are equal up to approximation errors r_i's that will be set smaller than estimation errors. Thus, the problem of estimating the parametric target $x_i'\beta_0$ is equivalent to the problem of estimating the non-parametric target f_i modulo approximation errors.

With that in mind, we choose our target or "true" β_0, with the corresponding cardinality of its support

$$s = \|\beta_0\|_0,$$

as any solution to the following ideal risk minimization or oracle problem:

$$\min_{\beta \in \mathbb{R}^p} \mathbb{E}_n[(f_i - x_i'\beta)^2] + \sigma^2 \frac{\|\beta\|_0}{n}. \tag{3.9}$$

We call this problem the oracle problem for the reasons explained below, and we call

$$T = \text{support}(\beta_0)$$

the oracle or the "true" model. Note that we necessarily have that $s \leqslant n$.

The oracle problem (3.9) balances the approximation error $\mathbb{E}_n[(f_i - x_i'\beta)^2]$ over the design points with the variance term $\sigma^2 \|\beta\|_0/n$, where the latter is determined by the number of non-zero coefficients in β. Letting

$$c_s^2 := \mathbb{E}_n[r_i^2] = \mathbb{E}_n[(f_i - x_i'\beta_0)^2]$$

denote the average square error from approximating values f_i by $x_i'\beta_0$, the quantity $c_s^2 + \sigma^2 s/n$ is the optimal value of (3.9). Typically, the optimality in (3.9) would balance the approximation error with the variance term so that for some absolute constant $K \geqslant 0$

$$c_s \leqslant K\sigma\sqrt{s/n}, \tag{3.10}$$

so that $\sqrt{c_s^2 + \sigma^2 s/n} \lesssim \sigma\sqrt{s/n}$. Thus, the quantity $\sigma\sqrt{s/n}$ becomes the ideal goal for the rate of convergence. If we knew the oracle model T, we would achieve this rate by using the oracle estimator, the least squares estimator based on this model, but we in general do not know T, since we do not observe the f_i's to attempt to solve the oracle problem (3.9). Since T is unknown, we will not be able to achieve the exact oracle rates of convergence, but we can hope to come close to this rate.

We consider the case of fixed design, namely we treat the covariate values $x_1, \ldots, x_n$ as fixed. This includes random sampling as a special case; indeed, in this case $x_1, \ldots, x_n$ represent a realization of this sample on which we condition throughout. Without loss of generality, we normalize the covariates so that

$$\widehat{\sigma}_j^2 = \mathbb{E}_n[x_{ij}^2] = 1 \text{ for } j = 1, \dots, p. \tag{3.11}$$

We summarize the setup as the following condition.

Condition ASM. *We have data* $\{(y_i, z_i), i = 1, \dots, n\}$ *that for each n obey the regression model (3.7), which admits the approximately sparse form (3.8) induced by (3.9) with the approximation error satisfying (3.10). The regressors* $x_i = P(z_i)$ *are normalized as in (3.11).*

Remark 3.1 (On the Oracle Problem). Let us now briefly explain what is behind problem (3.9). Under some mild assumptions, this problem directly arises as the (infeasible) oracle risk minimization problem. Indeed, consider an OLS estimator $\widehat{\beta}[\widetilde{T}]$, which is obtained by using a model $\widetilde{T}$, i.e. by regressing y_i on regressors $x_i[\widetilde{T}]$, where $x_i[\widetilde{T}] = \{x_{ij}, j \in \widetilde{T}\}$. This estimator takes value $\widehat{\beta}[\widetilde{T}] = \mathbb{E}_n[x_i[\widetilde{T}]x_i[\widetilde{T}]']^{-}\mathbb{E}_n[x_i[\widetilde{T}]y_i]$. The expected risk of this estimator $\mathbb{E}_n E[f_i - x_i[\widetilde{T}]'\widehat{\beta}[\widetilde{T}]]^2$ is equal to

$$\min_{\beta \in \mathbb{R}^{|\widetilde{T}|}} \mathbb{E}_n[(f_i - x_i[\widetilde{T}]'\beta)^2] + \sigma^2 \frac{k}{n},$$

where $k = \text{rank}(\mathbb{E}_n[x_i[\widetilde{T}]x_i[\widetilde{T}]'])$. The oracle knows the risk of each of the models $\widetilde{T}$ and can minimize this risk

$$\min_{\widetilde{T}} \min_{\beta \in \mathbb{R}^{|\widetilde{T}|}} \mathbb{E}_n[(f_i - x_i[\widetilde{T}]'\beta)^2] + \sigma^2 \frac{k}{n},$$

by choosing the best model or the oracle model T. This problem is in fact equivalent to (3.9), provided that rank $(\mathbb{E}_n[x_i[T]x_i[T]']) = \|\beta_0\|_0$, i.e. full rank. Thus, in this case the value β_0 solving (3.9) is the expected value of the oracle least squares estimator $\widehat{\beta}_I = \mathbb{E}_n[x_i[T]x_i[T]']^{-1}\mathbb{E}_n[x_i[T]y_i]$, i.e. $\beta_0 = \mathbb{E}_n[x_i[T]x_i[T]']^{-1}\mathbb{E}_n[x_i[T]f_i]$. This value is our target or "true" parameter value and the oracle model T is the target or "true" model. Note that when $c_s = 0$ we have that $f_i = x_i'\beta_0$, which gives us the special parametric case.

3.2.2 LASSO and Post-LASSO Estimators

Having introduced the model (3.8) with the target parameter defined via (3.9), our task becomes to estimate β_0. We will focus on deriving rate of convergence results in the *prediction norm*, which measures the accuracy of predicting $x_i'\beta_0$ over the design points $x_1, \dots, x_n$,

$$\|\delta\|_{2,n} = \sqrt{\mathbb{E}_n[x_i'\delta]^2}.$$

In what follows δ will denote deviations of the estimators from the true parameter value. Thus, e.g., for $\delta = \widehat{\beta} - \beta_0$, the quantity $\|\delta\|_{2,n}^2$ denotes the average of the square errors $x_i'\widehat{\beta} - x_i'\beta_0$ resulting from using the estimate $x_i'\widehat{\beta}$ instead of $x_i'\beta_0$. Note that once we bound $\widehat{\beta} - \beta_0$ in the prediction norm, we can also bound the empirical risk of predicting values f_i by $x_i'\beta$, via the triangle inequality:

$$\sqrt{\mathbb{E}_n[(x_i'\widehat{\beta} - f_i)^2]} \leqslant \|\widehat{\beta} - \beta_0\|_{2,n} + c_s. \tag{3.12}$$

In order to discuss estimation consider first the classical ideal AIC/BIC type estimator ([1, 22]) that solves the empirical (feasible) analog of the oracle problem:

$$\min_{\beta \in \mathbb{R}^p} \widehat{Q}(\beta) + \frac{\lambda}{n}\|\beta\|_0,$$

where $\widehat{Q}(\beta) = \mathbb{E}_n[(y_i - x_i'\beta)^2]$ and $\|\beta\|_0 = \sum_{j=1}^p 1\{|\beta_j| > 0\}$ is the ℓ_0-norm and λ is the penalty level. This estimator has very attractive theoretical properties, but unfortunately it is computationally prohibitive, since the solution to the problem may require solving $\sum_{k \leqslant n} \binom{p}{k}$ least squares problems (generically, the complexity of this problem is NP-hard [19, 12]).

One way to overcome the computational difficulty is to consider a convex relaxation of the preceding problem, namely to employ the closest convex penalty – the ℓ_1 penalty – in place of the ℓ_0 penalty. This construction leads to the so called LASSO estimator:[1]

$$\widehat{\beta} \in \arg\min_{\beta \in \mathbb{R}^p} \widehat{Q}(\beta) + \frac{\lambda}{n}\|\beta\|_1, \tag{3.13}$$

where as before $\widehat{Q}(\beta) = \mathbb{E}_n[(y_i - x_i'\beta)^2]$ and $\|\beta\|_1 = \sum_{j=1}^p |\beta_j|$. The LASSO estimator minimizes a convex function. Therefore, from a computational complexity perspective, (3.13) is a computationally efficient (i.e. solvable in polynomial time) alternative to AIC/BIC estimator.

In order to describe the choice of λ, we highlight that the following key quantity determining this choice:

$$S = 2\mathbb{E}_n[x_i\varepsilon_i],$$

which summarizes the noise in the problem. We would like to choose the smaller penalty level so that

$$\lambda \geqslant cn\|S\|_\infty \text{ with probability at least } 1 - \alpha, \tag{3.14}$$

where $1 - \alpha$ needs to be close to one, and c is a constant such that $c > 1$. Following [7] and [8], respectively, we consider two choices of λ that achieve the above:

$$X\text{-independent penalty:} \quad \lambda := 2c\sigma\sqrt{n}\Phi^{-1}(1 - \alpha/2p), \tag{3.15}$$

$$X\text{-dependent penalty:} \quad \lambda := 2c\sigma\Lambda(1 - \alpha|X), \tag{3.16}$$

where $\alpha \in (0, 1)$ and $c > 1$ is constant, and

---

[1] The abbreviation LASSO stands for Least Absolute Shrinkage and Selection Operator, c.f. [23].

$$\Lambda(1-\alpha|X) := (1-\alpha) - \text{quantile of } n\|S/(2\sigma)\|_\infty,$$

conditional on $X = (x_1, \ldots, x_n)'$. Note that

$$\|S/(2\sigma)\|_\infty =_d \max_{1 \leqslant j \leqslant p} |\mathbb{E}_n[x_{ij}g_i]|, \text{ where } g_i\text{'s are i.i.d. } N(0,1),$$

conditional on X, so we can compute $\Lambda(1-\alpha|X)$ simply by simulating the latter quantity, given the fixed design matrix X. Regarding the choice of α and c, asymptotically we require $\alpha \to 0$ as $n \to \infty$ and $c > 1$. Non-asymptotically, in our finite-sample experiments, $\alpha = .1$ and $c = 1.1$ work quite well. The noise level σ is unknown in practice, but we can estimate it consistently using the approach of Section 6. We recommend the X-dependent rule over the X-independent rule, since the former by construction adapts to the design matrix X and is less conservative than the latter in view of the following relationship that follows from Lemma 3.8:

$$\Lambda(1-\alpha|X) \leqslant \sqrt{n}\Phi^{-1}(1-\alpha/2p) \leqslant \sqrt{2n\log(2p/\alpha)}. \qquad (3.17)$$

Regularization by the ℓ_1-norm employed in (3.13) naturally helps the LASSO estimator to avoid overfitting the data, but it also shrinks the fitted coefficients towards zero, causing a potentially significant bias. In order to remove some of this bias, let us consider the Post-LASSO estimator that applies ordinary least squares regression to the model $\widehat{T}$ selected by LASSO. Formally, set

$$\widehat{T} = \text{support}(\widehat{\beta}) = \{j \in \{1, \ldots, p\} : |\widehat{\beta}_j| > 0\},$$

and define the Post-LASSO estimator $\widetilde{\beta}$ as

$$\widetilde{\beta} \in \arg\min_{\beta \in \mathbb{R}^p} \widehat{Q}(\beta) : \beta_j = 0 \text{ for each } j \in \widehat{T}^c, \qquad (3.18)$$

where $\widehat{T}^c = \{1, \ldots, p\} \setminus \widehat{T}$. In words, the estimator is ordinary least squares applied to the data after removing the regressors that were not selected by LASSO. If the model selection works perfectly – that is, $\widehat{T} = T$ – then the Post-LASSO estimator is simply the oracle estimator whose properties are well known. However, perfect model selection might be unlikely for many designs of interest, so we are especially interested in the properties of Post-LASSO in such cases, namely when $\widehat{T} \neq T$, especially when $T \not\subseteq \widehat{T}$.

3.2.3 Intuition and Geometry of LASSO and Post-LASSO

In this section we discuss the intuition behind LASSO and Post-LASSO estimators defined above. We shall rely on a dual interpretation of the LASSO optimization problem to provide some geometrical intuition for the performance of LASSO.

Indeed, it can be seen that the LASSO estimator also solves the following optimization program:

$$\min_{\beta \in \mathbb{R}^p} \|\beta\|_1 : \widehat{Q}(\beta) \leqslant \gamma \tag{3.19}$$

for some value of $\gamma \geqslant 0$ (that depends on the penalty level λ). Thus, the estimator minimizes the ℓ_1-norm of coefficients subject to maintaining a certain goodness-of-fit; or, geometrically, the LASSO estimator searches for a minimal ℓ_1-ball – the diamond– subject to the diamond having a non-empty intersection with a fixed lower contour set of the least squares criterion function – the ellipse.

In Figure 3.2 we show an illustration for the two-dimensional case with the true parameter value (β_{01}, β_{02}) equal $(1, 0)$, so that $T = \text{support}(\beta_0) = \{1\}$ and $s = 1$. In the figure we plot the diamonds and ellipses. In the top figure, the ellipse represents a lower contour set of the population criterion function $Q(\beta) = E[(y_i - x_i'\beta)^2]$ in the zero noise case or the infinite sample case. In the bottom figures the ellipse represents a contour set of the sample criterion function $\widehat{Q}(\beta) = \mathbb{E}_n[(y_i - x_i'\beta)^2]$ in the non-zero noise or the finite sample case. The set of optimal solutions $\widehat{\beta}$ for LASSO is then given by the intersection of the minimal diamonds with the ellipses. Finally, recall that Post-LASSO is computed as the ordinary least square solution using covariates selected by LASSO. Thus, Post-LASSO estimate $\widetilde{\beta}$ is given by the center of the ellipse intersected with the linear subspace selected by LASSO.

In the zero-noise case or in population (top figure), LASSO easily recovers the correct sparsity pattern of β_0. Note that due to the regularization, in spite of the absence of noise, the LASSO estimator has a large bias towards zero. However, in this case Post-LASSO $\widetilde{\beta}$ removes the bias and recovers β_0 perfectly.

In the non-zero noise case (middle and bottom figures), the contours of the criterion function and its center move away from the population counterpart. The empirical error in the middle figure moves the center of the ellipse to a non-sparse point. However, LASSO correctly sets $\widehat{\beta}_2 = 0$ and $\widehat{\beta}_1 \neq 0$ recovering the sparsity pattern of β_0. Using the selected support, Post-LASSO $\widetilde{\beta}$ becomes the oracle estimator which drastically improves upon LASSO. In the case of the bottom figure, we have large empirical errors that push the center of the lower contour set further away from the population counterpart. These large empirical errors make the LASSO estimator non-sparse, incorrectly setting $\widehat{\beta}_2 \neq 0$. Therefore, Post-LASSO uses $\widehat{T} = \{1, 2\}$ and does not use the exact support $T = \{1\}$. Thus, Post-LASSO is not the oracle estimator in this case.

All three figures also illustrate the shrinkage bias towards zero in the LASSO estimator that is introduced by the ℓ_1-norm penalty. The Post-LASSO estimator is motivated as a solution to remove (or at least alleviate) this shrinkage bias. In cases where LASSO achieves a good sparsity pattern, Post-LASSO can drastically improve upon LASSO.

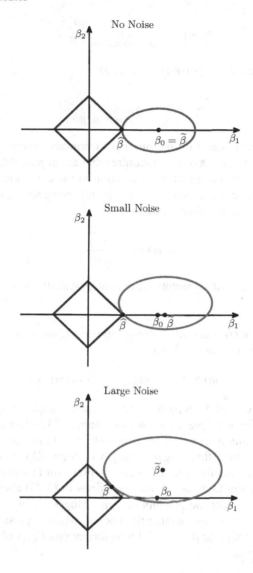

Fig. 3.2 The figures illustrate the geometry of LASSO and Post-LASSO estimator.

3.2.4 Primitive Conditions

In both the parametric and non-parametric models described above, whenever $p > n$, the empirical Gram matrix $\mathbb{E}_n[x_i x_i']$ does not have full rank and hence it is not well-behaved. However, we only need good behavior of certain moduli of continuity of the Gram matrix called restricted sparse eigenvalues. We define the minimal restricted sparse eigenvalue

$$\kappa(m)^2 := \min_{\|\delta_{T^c}\|_0 \leqslant m, \delta \neq 0} \frac{\|\delta\|_{2,n}^2}{\|\delta\|^2}, \tag{3.20}$$

and the maximal restricted sparse eigenvalue as

$$\phi(m) := \max_{\|\delta_{T^c}\|_0 \leqslant m, \delta \neq 0} \frac{\|\delta\|_{2,n}^2}{\|\delta\|^2}, \tag{3.21}$$

where m is the upper bound on the number of non-zero components outside the support T. To assume that $\kappa(m) > 0$ requires that all empirical Gram submatrices formed by any m components of x_i in addition to the components in T are positive definite. It will be convenient to define the following sparse *condition* number associated with the empirical Gram matrix:

$$\mu(m) = \frac{\sqrt{\phi(m)}}{\kappa(m)}. \tag{3.22}$$

In order to state simplified asymptotic statements, we shall also invoke the following condition.

Condition RSE. *Sparse eigenvalues of the empirical Gram matrix are well behaved, in the sense that for $m = m_n = s \log n$*

$$\mu(m) \lesssim 1, \quad \phi(m) \lesssim 1, \quad 1/\kappa(m) \lesssim 1. \tag{3.23}$$

This condition holds with high probability for many designs of interest under mild conditions on s. For example, as shown in Lemma 3.1, when the covariates are Gaussians, the conditions in (3.23) are true with probability converging to one under the mild assumption that $s \log p = o(n)$. Condition RSE is likely to hold for other regressors with jointly light-tailed distributions, for instance log-concave distribution. As shown in Lemma 3.2, the conditions in (3.23) also hold for general bounded regressors under the assumption that $s^2 \log p = o(n)$. Arbitrary bounded regressors often arise in non-parametric models, where regressors x_i are formed as spline, trigonometric, or polynomial transformations $P(z_i)$ of some elementary bounded regressors z_i.

Lemma 3.1 (Gaussian design). *Suppose $\tilde{x}_i$, $i = 1, \ldots, n$, are i.i.d. zero-mean Gaussian random vectors, such that the population design matrix $E[\tilde{x}_i \tilde{x}_i']$ has ones on the diagonal, and its eigenvalues are bounded from above by $\varphi < \infty$ and bounded from below by $\kappa^2 > 0$. Define x_i as a normalized form of $\tilde{x}_i$, namely $x_{ij} = \tilde{x}_{ij}/\sqrt{\mathbb{E}_n[\tilde{x}_{ij}^2]}$. Then for any $m \leqslant (s \log(n/e)) \wedge (n/[16 \log p])$, with probability at least $1 - 2\exp(-n/16)$,*

$$\phi(m) \leqslant 8\varphi, \quad \kappa(m) \geqslant \kappa/6\sqrt{2}, \quad \text{and} \quad \mu(m) \leqslant 24\sqrt{\varphi}/\kappa.$$

Lemma 3.2 (Bounded design). *Suppose* $\tilde{x}_i \ i = 1, \ldots, n,$ *are i.i.d. bounded zero-mean random vectors, with* $\max_{1 \leqslant i \leqslant n, 1 \leqslant j \leqslant p} |\tilde{x}_{ij}| \leqslant K_n$ *with probability 1 for all n. Assume that the population design matrix* $E[\tilde{x}_i \tilde{x}_i']$ *has ones on the diagonal, and its eigenvalues are bounded from above by* $\varphi < \infty$ *and bounded from below by* $\kappa^2 > 0$. *Define* x_i *as a normalized form of* $\tilde{x}_i$, *namely* $x_{ij} = \tilde{x}_{ij} / \sqrt{\mathbb{E}_n[\tilde{x}_{ij}^2]}$. *Then there is a constant* $\bar{\varepsilon} > 0$ *such that if* $\sqrt{n}/K_n \to \infty$ *and* $m \leqslant (s \log(n/e)) \wedge ([\bar{\varepsilon}/K_n] \sqrt{n/\log p})$, *we have that as* $n \to \infty$

$$\phi(m) \leqslant 4\varphi, \quad \kappa(m) \geqslant \kappa/2, \quad \text{and} \quad \mu(m) \leqslant 4\sqrt{\varphi}/\kappa,$$

with probability approaching 1.

For proofs, see [7]; both lemmas build upon results in [25].

3.3 Analysis of LASSO

In this section we discuss the rate of convergence of LASSO in the prediction norm; our exposition follows mainly [8].

The key quantity in the analysis is the following quantity called "score":

$$S = S(\beta_0) = 2\mathbb{E}_n[x_i \varepsilon_i].$$

The score is the effective "noise" in the problem. Indeed, defining $\delta := \hat{\beta} - \beta_0$, note that by the Hölder's inequality

$$\begin{aligned}
\hat{Q}(\hat{\beta}) - \hat{Q}(\beta_0) - \|\delta\|_{2,n}^2 &= -2\mathbb{E}_n[\varepsilon_i x_i' \delta] - 2\mathbb{E}_n[r_i x_i' \delta] \\
&\geqslant -\|S\|_\infty \|\delta\|_1 - 2c_s \|\delta\|_{2,n}.
\end{aligned} \tag{3.24}$$

Intuition suggests that we need to majorize the "noise term" $\|S\|_\infty$ by the penalty level λ/n, so that the bound on $\|\delta\|_{2,n}^2$ will follow from a relation between the prediction norm $\|\cdot\|_{2,n}$ and the penalization norm $\|\cdot\|_1$ on a suitable set. Specifically, for any $c > 1$, it will follow that if

$$\lambda \geqslant cn\|S\|_\infty$$

and $\|\delta\|_{2,n} \geqslant 2c_s$, the vector δ will also satisfy

$$\|\delta_{T^c}\|_1 \leqslant \bar{c}\|\delta_T\|_1, \tag{3.25}$$

where $\bar{c} = (c+1)/(c-1)$. That is, in this case the error in the regularization norm outside the true support does not exceed $\bar{c}$ times the error in the true support. (In the case $\|\delta\|_{2,n} \leqslant 2c_s$ the inequality (3.25) may not hold, but the bound $\|\delta\|_{2,n} \leqslant 2c_s$ is already good enough.)

Consequently, the analysis of the rate of convergence of LASSO relies on the so-called restricted eigenvalue $\kappa_{\bar{c}}$ introduced in [8], which controls the modulus of continuity between the prediction norm $\|\cdot\|_{2,n}$ and the penalization norm $\|\cdot\|_1$ over the set of vectors $\delta \in \mathbb{R}^p$ that satisfy (3.25):

$$\kappa_{\bar{c}} := \min_{\|\delta_{T^c}\|_1 \leqslant \bar{c}\|\delta_T\|_1, \delta_T \neq 0} \frac{\sqrt{s}\|\delta\|_{2,n}}{\|\delta_T\|_1}, \qquad (\mathrm{RE}(c))$$

where $\kappa_{\bar{c}}$ can depend on n. The constant $\kappa_{\bar{c}}$ is a crucial technical quantity in our analysis and we need to bound it away from zero. In the leading cases that condition RSE holds this will in fact be the case as the sample size grows, namely

$$1/\kappa_{\bar{c}} \lesssim 1. \qquad (3.26)$$

Indeed, we can bound $\kappa_{\bar{c}}$ from below by

$$\kappa_{\bar{c}} \geqslant \max_{m \geqslant 0} \kappa(m) \left(1 - \mu(m)\,\bar{c}\sqrt{s/m}\right) \geqslant \kappa(s \log n)\left(1 - \mu(s \log n)\,\bar{c}\sqrt{1/\log n}\right)$$

by Lemma 3.10 stated and proved in the appendix. Thus, under the condition RSE, as n grows, $\kappa_{\bar{c}}$ is bounded away from zero since $\kappa(s \log n)$ is bounded away from zero and $\phi(s \log n)$ is bounded from above as in (3.23). Several other primitive assumptions can be used to bound $\kappa_{\bar{c}}$. We refer the reader to [8] for a further detailed discussion of lower bounds on $\kappa_{\bar{c}}$.

We next state a non-asymptotic performance bound for the LASSO estimator.

Theorem 3.1 (Non-Asymptotic Bound for LASSO). *Under condition ASM, the event $\lambda \geqslant cn\|S\|_\infty$ implies*

$$\|\widehat{\beta} - \beta_0\|_{2,n} \leqslant \left(1 + \frac{1}{c}\right)\frac{\lambda\sqrt{s}}{n\kappa_{\bar{c}}} + 2c_s, \qquad (3.27)$$

where $c_s = 0$ in the parametric case, and $\bar{c} = (c+1)/(c-1)$. Thus, if $\lambda \geqslant cn\|S\|_\infty$ with probability at least $1 - \alpha$, as guaranteed by either X-independent or X-dependent penalty levels (3.15) and (3.15), then the bound (3.27) occurs with probability at least $1 - \alpha$.

The proof of Theorem 3.1 is given in the appendix. The theorem also leads to the following useful asymptotic bounds.

Corollary 3.1 (Asymptotic Bound for LASSO). *Suppose that conditions ASM and RSE hold. If λ is chosen according to either the X-independent or X-dependent rule specified in (3.15) and (3.16) with $\alpha = o(1)$, $\log(1/\alpha) \lesssim \log p$, or more generally so that*

$$\lambda \lesssim_P \sigma\sqrt{n \log p} \quad \text{and} \quad \lambda \geqslant c'n\|S\|_\infty \, wp \rightarrow 1, \qquad (3.28)$$

for some $c' > 1$, then the following asymptotic bound holds:

$$\|\widehat{\beta} - \beta_0\|_{2,n} \lesssim_P \sigma\sqrt{\frac{s \log p}{n}} + c_s.$$

The non-asymptotic and asymptotic bounds for the empirical risk immediately follow from the triangle inequality:

$$\sqrt{\mathbb{E}_n[(f_i - x_i'\widehat{\beta})^2]} \leqslant \|\widehat{\beta} - \beta_0\|_{2,n} + c_s. \tag{3.29}$$

Thus, the rate of convergence of $x_i'\widehat{\beta}$ to f_i coincides with the rate of convergence of the oracle estimator $\sqrt{c_s^2 + \sigma^2 s/n}$ up to a logarithmic factor of p. Nonetheless, the performance of LASSO can be considered optimal in the sense that under general conditions the oracle rate is achievable only up to logarithmic factor of p (see Donoho and Johnstone [11] and Rigollet and Tsybakov [20]), apart from very exceptional, stringent cases, in which it is possible to perform perfect or near-perfect model selection.

3.4 Model Selection Properties and Sparsity of LASSO

The purpose of this section is, first, to provide bounds (sparsity bounds) on the dimension of the model selected by LASSO, and, second, to describe some special cases where the model selected by LASSO perfectly matches the "true" (oracle) model.

3.4.1 Sparsity Bounds

Although perfect model selection can be seen as unlikely in many designs, sparsity of the LASSO estimator has been documented in a variety of designs. Here we describe the sparsity results obtained in [7]. Let us define

$$\widehat{m} := |\widehat{T} \setminus T| = \|\widehat{\beta}_{T^c}\|_0,$$

which is the number of unnecessary regressors selected by LASSO.

Theorem 3.2 (Non-Asymptotic Sparsity Bound for LASSO). *Suppose condition ASM holds. The event* $\lambda \geqslant cn\|S\|_\infty$ *implies that*

$$\widehat{m} \leqslant s \cdot \left[\min_{m \in \mathscr{M}} \phi(m \wedge n) \right] \cdot L,$$

where $\mathscr{M} = \{m \in \mathbb{N} : m > s\phi(m \wedge n) \cdot 2L\}$ *and* $L = [2\overline{c}/\kappa_{\overline{c}} + 3(\overline{c}+1)nc_s/(\lambda\sqrt{s})]^2$.
Under Conditions ASM and RSE, for n sufficiently large we have $1/\kappa_{\overline{c}} \lesssim 1$, $c_s \lesssim \sigma\sqrt{s/n}$, and $\phi(s\log n) \lesssim 1$; and under the conditions of Corollary 3.1, $\lambda \geqslant c\sigma\sqrt{n}$ with probability approaching one. Therefore, we have that $L \lesssim_P 1$ and

$$s\log n > s\phi(s\log n) \cdot 2L, \quad \text{that is,} \quad s\log n \in \mathscr{M}$$

with probability approaching one as n grows. Therefore, under these conditions we have

$$\min_{m \in \mathcal{M}} \phi(m \wedge n) \lesssim_P 1.$$

Corollary 3.2 (Asymptotic Sparsity Bound for LASSO). *Under the conditions of Corollary 3.1, we have that*

$$\widehat{m} \lesssim_P s. \tag{3.30}$$

Thus, using a penalty level that satisfies (3.28) LASSO's sparsity is asymptotically of the same order as the oracle sparsity, namely

$$\widehat{s} := |\widehat{T}| \leqslant s + \widehat{m} \lesssim_P s. \tag{3.31}$$

We note here that Theorem 3.2 is particularly helpful in designs in which $\min_{m \in \mathcal{M}} \phi(m) \ll \phi(n)$. This allows Theorem 3.2 to sharpen the sparsity bound of the form $\widehat{s} \lesssim_P s\phi(n)$ considered in [8] and [18]. The bound above is comparable to the bounds in [25] in terms of order of magnitude, but Theorem 3.2 requires a smaller penalty level λ which also does not depend on the unknown sparse eigenvalues as in [25].

3.4.2 Perfect Model Selection Results

The purpose of this section is to describe very special cases where perfect model selection is possible. Most results in the literature for model selection have been developed for the parametric case only ([18],[17]). Below we provide some results for the nonparametric models, which cover the parametric models as a special case.

Lemma 3.3 (Cases with Perfect Model Selection by Thresholded LASSO). *Suppose condition ASM holds. (1) If the non-zero coefficients of the oracle model are well separated from zero, that is*

$$\min_{j \in T} |\beta_{0j}| > \zeta + t, \quad \text{for some } t \geqslant \zeta := \max_{j=1,\dots,p} |\widehat{\beta}_j - \beta_{0j}|,$$

then the oracle model is a subset of the selected model,

$$T := \text{support}(\beta_0) \subseteq \widehat{T} := \text{support}(\widehat{\beta}).$$

Moreover the oracle model T can be perfectly selected by applying hard-thresholding of level t to the LASSO estimator $\widehat{\beta}$:

$$T = \left\{ j \in \{1,\dots,p\} : |\widehat{\beta}_j| > t \right\}.$$

(2) In particular, if $\lambda \geqslant cn\|S\|_\infty$, then for $\widehat{m} = |\widehat{T} \setminus T| = \|\widehat{\beta}_{T^c}\|_0$ we have

$$\zeta \leqslant \left(1+\frac{1}{c}\right)\frac{\lambda\sqrt{s}}{n\kappa_{\overline{c}}\kappa(\widehat{m})}+\frac{2c_s}{\kappa(\widehat{m})}.$$

(3) In particular, if $\lambda \geqslant cn\|S\|_\infty$, and there is a constant $U > 5\overline{c}$ such that the empirical Gram matrix satisfies $|\mathbb{E}_n[x_{ij}x_{ik}]| \leqslant 1/[Us]$ for all $1 \leqslant j < k \leqslant p$, then

$$\zeta \leqslant \frac{\lambda}{n}\cdot\frac{U+\overline{c}}{U-5\overline{c}}+\min\left\{\frac{\sigma}{\sqrt{n}},c_s\right\}+\frac{6\overline{c}}{U-5\overline{c}}\frac{c_s}{\sqrt{s}}+\frac{4\overline{c}n}{U}\frac{c_s^2}{\lambda}\frac{c_s^2}{s}.$$

Thus, we see from parts (1) and (2) that perfect model selection is possible under strong assumptions on the coefficients' separation away from zero. We also see from part (3) that the strong separation of coefficients can be considerably weakened in exchange for a strong assumption on the maximal pairwise correlation of regressors. These results generalize to the nonparametric case the results of [17] and [18] for the parametric case in which $c_s = 0$.

Finally, the following result on perfect model selection also requires strong assumptions on separation of coefficients and the empirical Gram matrix. Recall that for a scalar v, $\text{sign}(v) = v/|v|$ if $|v| > 0$, and 0 otherwise. If v is a vector, we apply the definition componentwise. Also, given a vector $x \in \mathbb{R}^p$ and a set $T \subset \{1,...,p\}$, let us denote $x_i[T] := \{x_{ij}, j \in T\}$.

Lemma 3.4 (Cases with Perfect Model Selection by LASSO). *Suppose condition ASM holds. We have perfect model selection for LASSO, $\widehat{T} = T$, if and only if*

$$\left\|\mathbb{E}_n[x_i[T^c]x_i[T]']\mathbb{E}_n[x_i[T]x_i[T]']^{-1}\left\{\mathbb{E}_n[x_i[T]u_i]\right.\right.$$

$$\left.\left.-\frac{\lambda}{2n}\text{sign}(\beta_0[T])\right\}-\mathbb{E}_n[x_i[T^c]u_i]\right\|_\infty \leqslant \frac{\lambda}{2n},$$

$$\min_{j\in T}\left|\beta_{0j}+\left(\mathbb{E}_n[x_i[T]x_i[T]']^{-1}\left\{\mathbb{E}_n[x_i[T]u_i]-\frac{\lambda}{2n}\text{sign}(\beta_0[T])\right\}\right)_j\right| > 0.$$

The result follows immediately from the first order optimality conditions, see [24]. [26] and [9] provides further primitive sufficient conditions for perfect model selection for the parametric case in which $u_i = \varepsilon_i$. The conditions above might typically require a slightly larger choice of λ than (3.15) and larger separation from zero of the minimal non-zero coefficient $\min_{j\in T}|\beta_{0j}|$.

3.5 Analysis of Post-LASSO

Next we study the rate of convergence of the Post-LASSO estimator. Recall that for $\widehat{T} = \text{support}(\widehat{\beta})$, the Post-LASSO estimator solves

$$\widetilde{\beta} \in \arg\min_{\beta\in\mathbb{R}^p} \widehat{Q}(\beta) : \beta_j = 0 \text{ for each } j \in \widehat{T}^c.$$

It is clear that if the model selection works perfectly (as it will under some rather stringent conditions discussed in Section 3.4.2), that is, $T = \widehat{T}$, then this estimator is simply the oracle least squares estimator whose properties are well known. However, if the model selection does not work perfectly, that is, $T \neq \widehat{T}$, the resulting performance of the estimator faces two different perils: First, in the case where LASSO selects a model $\widehat{T}$ that does not fully include the true model T, we have a specification error in the second step. Second, if LASSO includes additional regressors outside T, these regressors were not chosen at random and are likely to be spuriously correlated with the disturbances, so we have a data-snooping bias in the second step.

It turns out that despite of the possible poor selection of the model, and the aforementioned perils this causes, the Post-LASSO estimator still performs well theoretically, as shown in [7]. Here we provide a proof similar to [6] which is easier generalize to non-Gaussian cases.

Theorem 3.3 (Non-Asymptotic Bound for Post-LASSO). *Suppose condition ASM holds. If $\lambda \geqslant cn\|S\|_\infty$ holds with probability at least $1 - \alpha$, then for any $\gamma > 0$ there is a constant K_γ independent of n such that with probability at least $1 - \alpha - \gamma$*

$$\|\widetilde{\beta} - \beta_0\|_{2,n} \leqslant \frac{K_\gamma \sigma}{\kappa(\widehat{m})}\sqrt{\frac{s + \widehat{m}\log p}{n}} + 2c_s + 1\{T \not\subseteq \widehat{T}\}\sqrt{\frac{\lambda\sqrt{s}}{n\kappa_{\bar{c}}} \cdot \left(\frac{(1+c)\lambda\sqrt{s}}{cn\kappa_{\bar{c}}} + 2c_s\right)}.$$

This theorem provides a performance bound for Post-LASSO as a function of LASSO's sparsity characterized by $\widehat{m}$, LASSO's rate of convergence, and LASSO's model selection ability. For common designs this bound implies that Post-LASSO performs at least as well as LASSO, but it can be strictly better in some cases, and has a smaller shrinkage bias by construction.

Corollary 3.3 (Asymptotic Bound for Post-LASSO). *Suppose conditions of Corollary 3.1 hold. Then*

$$\|\widetilde{\beta} - \beta_0\|_{2,n} \lesssim_P \sigma\sqrt{\frac{s\log p}{n}} + c_s. \tag{3.32}$$

If further $\widehat{m} = o(s)$ and $T \subseteq \widehat{T}$ with probability approaching one, then

$$\|\widetilde{\beta} - \beta_0\|_{2,n} \lesssim_P \sigma\left[\sqrt{\frac{o(s)\log p}{n}} + \sqrt{\frac{s}{n}}\right] + c_s. \tag{3.33}$$

If $\widehat{T} = T$ with probability approaching one, then Post-LASSO achieves the oracle performance

$$\|\widetilde{\beta} - \beta_0\|_{2,n} \lesssim_P \sigma\sqrt{s/n} + c_s. \tag{3.34}$$

It is also worth repeating here that finite-sample and asymptotic bounds in other norms of interest immediately follow by the triangle inequality and by definition of $\kappa(\widehat{m})$:

$$\sqrt{\mathbb{E}_n[(x_i'\tilde{\beta} - f_i)^2]} \leq \|\tilde{\beta} - \beta_0\|_{2,n} + c_s \text{ and } \|\tilde{\beta} - \beta_0\| \leq \|\tilde{\beta} - \beta_0\|_{2,n}/\kappa(\widehat{m}).$$

The corollary above shows that Post-LASSO achieves the same near-oracle rate as LASSO. Notably, this occurs despite the fact that LASSO may in general fail to correctly select the oracle model T as a subset, that is $T \not\subseteq \widehat{T}$. The intuition for this result is that any components of T that LASSO misses cannot be very important. This corollary also shows that in some special cases Post-LASSO strictly improves upon LASSO's rate. Finally, note that Corollary 3.3 follows by observing that under the stated conditions,

$$\|\tilde{\beta} - \beta_0\|_{2,n} \lesssim_P \sigma\left[\sqrt{\frac{\widehat{m}\log p}{n}} + \sqrt{\frac{s}{n}} + 1\{T \not\subseteq \widehat{T}\}\sqrt{\frac{s\log p}{n}}\right] + c_s.$$

3.6 Estimation of Noise Level

Our specification of penalty levels (3.16) and (3.15) require the practitioner to know the noise level σ of the disturbances or at least estimate it. The purpose of this section is to propose the following method for estimating σ. First, we use a conservative estimate $\widehat{\sigma}^0 = \sqrt{\text{Var}_n[y_i]} := \sqrt{\mathbb{E}_n[(y_i - \bar{y})^2]}$, where $\bar{y} = \mathbb{E}_n[y_i]$, in place of σ^2 to obtain the initial LASSO and Post-LASSO estimates, $\widehat{\beta}$ and $\tilde{\beta}$. The estimate $\widehat{\sigma}^0$ is conservative since $\widehat{\sigma}^0 = \sigma^0 + o_P(1)$ where $\sigma^0 = \sqrt{\text{Var}[y_i]} \geq \sigma$, since x_i contains a constant by assumption. Second, we define the refined estimate $\widehat{\sigma}$ as

$$\widehat{\sigma} = \sqrt{\widehat{Q}(\widehat{\beta})}$$

in the case of LASSO and

$$\widehat{\sigma} = \sqrt{\frac{n}{n-\widehat{s}} \cdot \widehat{Q}(\tilde{\beta})}$$

in the case of Post-LASSO. In the latter case we employ the standard degree-of-freedom correction with $\widehat{s} = \|\tilde{\beta}\|_0 = |\widehat{T}|$, and in the former case we need no additional corrections, since the LASSO estimate is already sufficiently regularized. Third, we use the refined estimate $\widehat{\sigma}^2$ to obtain the refined LASSO and Post-LASSO estimates $\widehat{\beta}$ and $\tilde{\beta}$. We can stop here or further iterate on the last two steps.

Thus, the algorithm for estimating σ using LASSO is as follows:

Algorithm 1 (Estimation of σ using LASSO iterations) *Set $\widehat{\sigma}^0 = \sqrt{\text{Var}_n[y_i]}$ and $k = 0$, and specify a small constant $v > 0$, the tolerance level, and a constant $I > 1$, the upper bound on the number of iterations. (1) Compute the LASSO estimator $\widehat{\beta}$ based on $\lambda = 2c\widehat{\sigma}^k\Lambda(1 - \alpha|X)$. (2) Set*

$$\widehat{\sigma}^{k+1} = \sqrt{\widehat{Q}(\widehat{\beta})}.$$

(3) If $|\widehat{\sigma}^{k+1} - \widehat{\sigma}^k| \leqslant v$ or $k+1 \geqslant I$, then stop and report $\widehat{\sigma} = \widehat{\sigma}^{k+1}$; otherwise set $k \leftarrow k+1$ and go to (1).

And the algorithm for estimating σ using Post-LASSO is as follows:

Algorithm 2 (Estimation of σ using Post-LASSO iterations) *Set $\widehat{\sigma}^0 = \sqrt{Var_n[y_i]}$ and $k = 0$, and specify a small constant $v \geqslant 0$, the tolerance level, and a constant $I > 1$, the upper bound on the number of iterations. (1) Compute the Post-LASSO estimator $\widetilde{\beta}$ based on $\lambda = 2c\widehat{\sigma}^k\Lambda(1 - \alpha|X)$. (2) Set*

$$\widehat{\sigma}^{k+1} = \sqrt{\frac{n}{n-\widehat{s}} \cdot \widehat{Q}(\widetilde{\beta})},$$

where $\widehat{s} = \|\widetilde{\beta}\|_0 = |\widehat{T}|$. (3) If $|\widehat{\sigma}^{k+1} - \widehat{\sigma}^k| \leqslant v$ or $k+1 \geqslant I$, then stop and report $\widehat{\sigma} = \widehat{\sigma}^{k+1}$; otherwise, set $k \leftarrow k+1$ and go to (1).

We can also use $\lambda = 2c\widehat{\sigma}^k\sqrt{n}\Phi^{-1}(1 - \alpha/2p)$ in place of X-dependent penalty. We note that using LASSO to estimate σ it follows that the sequence $\widehat{\sigma}^k$, $k \geqslant 2$, is monotone, while using Post-LASSO the estimates $\widehat{\sigma}^k$, $k \geqslant 1$, can only assume a finite number of different values.

The following theorem shows that these algorithms produce consistent estimates of the noise level, and that the LASSO and Post-LASSO estimators based on the resulting data-driven penalty continue to obey the asymptotic bounds we have derived previously.

Theorem 3.4 (Validity of Results with Estimated σ). *Suppose conditions ASM and RES hold. Suppose that $\sigma \leqslant \widehat{\sigma}^0 \lesssim \sigma$ with probability approaching 1 and $s\log p/n \to 0$. Then $\widehat{\sigma}$ produced by either Algorithm 1 or 2 is consistent*

$$\widehat{\sigma}/\sigma \to_P 1$$

so that the penalty levels $\lambda = 2c\widehat{\sigma}^k\Lambda(1 - \alpha|X)$ and $\lambda = 2c\widehat{\sigma}^k\sqrt{n}\Phi^{-1}(1 - \alpha/2p)$ with $\alpha = o(1)$, and $\log(1/\alpha) \lesssim \log p$, satisfy the condition (3.28) of Corollary 1, namely

$$\lambda \lesssim_P \sigma\sqrt{n\log p} \quad \text{and} \quad \lambda \geqslant c'n\|S\|_\infty \, wp \to 1, \tag{3.35}$$

for some $1 < c' < c$. Consequently, the LASSO and Post-LASSO estimators based on this penalty level obey the conclusions of Corollaries 1, 2, and 3.

3.7 Monte Carlo Experiments

In this section we compare the performance of LASSO, Post-LASSO, and the ideal oracle linear regression estimators. The oracle estimator applies ordinary least square to the true model. (Such an estimator is not available outside Monte Carlo experiments.)

We begin by considering the following regression model:

$$y = x'\beta_0 + \varepsilon, \quad \beta_0 = (1, 1, 1/2, 1/3, 1/4, 1/5, 0, \ldots, 0)',$$

where $x = (1, z')'$ consists of an intercept and covariates $z \sim N(0, \Sigma)$, and the errors ε are independently and identically distributed $\varepsilon \sim N(0, \sigma^2)$. The dimension p of the covariates x is 500, the dimension s of the true model is 6, and the sample size n is 100. We set λ according to the X-dependent rule with $1 - \alpha = 90\%$. The regressors are correlated with $\Sigma_{ij} = \rho^{|i-j|}$ and $\rho = 0.5$. We consider two levels of noise: Design 1 with $\sigma^2 = 1$ (higher level) and Design 2 with $\sigma^2 = 0.1$ (lower level). For each repetition we draw new vectors x_i's and errors ε_i's.

We summarize the model selection performance of LASSO in Figures 3.3 and 3.4. In the left panels of the figures, we plot the frequencies of the dimensions of the selected model; in the right panels we plot the frequencies of selecting the correct regressors. From the left panels we see that the frequency of selecting a much larger model than the true model is very small in both designs. In the design with a larger noise, as the right panel of Figure 3.3 shows, LASSO frequently fails to select the entire true model, missing the regressors with small coefficients. However, it almost always includes the most important three regressors with the largest coefficients. Notably, despite this partial failure of the model selection Post-LASSO still performs well, as we report below. On the other hand, we see from the right panel of Figure 3.4 that in the design with a lower noise level LASSO rarely misses any component of the true support. These results confirm the theoretical results that when the non-zero coefficients are well-separated from zero, the penalized estimator should select a model that includes the true model as a subset. Moreover, these results also confirm the theoretical result of Theorem 3.2, namely, that the dimension of the selected model should be of the same stochastic order as the dimension of the true model. In summary, the model selection performance of the penalized estimator agrees very well with the theoretical results.

We summarize the results on the performance of estimators in Table 3.3, which records for each estimator $\check{\beta}$ the mean ℓ_0-norm $E[\|\check{\beta}\|_0]$, the norm of the bias $\|E\check{\beta} - \beta_0\|$ and also the prediction error $E[\mathbb{E}_n[|x_i'(\check{\beta} - \beta_0)|^2]^{1/2}]$ for recovering the regression function. As expected, LASSO has a substantial bias. We see that Post-LASSO drastically improves upon the LASSO, particularly in terms of reducing the bias, which also results in a much lower overall prediction error. Notably, despite that under the higher noise level LASSO frequently fails to recover the true model, the Post-LASSO estimator still performs well. This is because the penalized estimator always manages to select the most important regressors. We also see that the prediction error of the Post-LASSO is within a factor $\sqrt{\log p}$ of the prediction error of the oracle estimator, as we would expect from our theoretical results. Under the lower noise level, Post-LASSO performs almost identically to the ideal oracle estimator. We would expect this since in this case LASSO selects the model especially well making Post-LASSO nearly the oracle.

The results above used the true value of σ in the choice of λ. Next we illustrate how σ can be estimated in practice. We follow the iterative procedure described in the previous section. In our experiments the tolerance was 10^{-8} times the current estimate for σ, which is typically achieved in less than 15 iterations.

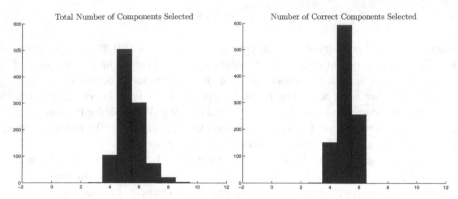

Fig. 3.3 The figure summarizes the covariate selection results for the design with $\sigma = 1$, based on 1000 Monte Carlo repetitions. The left panel plots the histogram for the number of covariates selected by LASSO out of the possible 500 covariates, $|\widehat{T}|$. The right panel plots the histogram for the number of significant covariates selected by LASSO, $|\widehat{T} \cap T|$; there are in total 6 significant covariates amongst 500 covariates. The sample size for each repetition was $n = 100$.

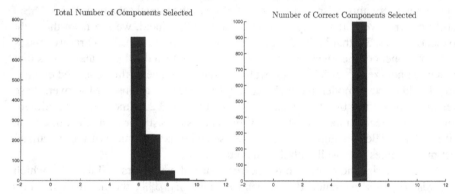

Fig. 3.4 The figure summarizes the covariate selection results for the design with $\sigma^2 = 0.1$, based on 1000 Monte Carlo repetitions. The left panel plots the histogram for the number of covariates selected out of the possible 500 covariates, $|\widehat{T}|$. The right panel plots the histogram for the number of significant covariates selected, $|\widehat{T} \cap T|$; there are in total 6 significant covariates amongst 500 covariates. The sample size for each repetition was $n = 100$.

We assess the performance of the iterative procedure under the design with the larger noise, $\sigma^2 = 1$ (similar results hold for $\sigma^2 = 0.1$). The histograms in Figure 3.5 show that the model selection properties are very similar to the model selection when σ is known. Figure 3.6 displays the distribution of the estimator $\widehat{\sigma}$ of σ based on (iterative) Post-LASSO, (iterative) LASSO, and the initial estimator $\widehat{\sigma}^0 = \sqrt{\mathrm{Var}_n[y_i]}$. As we expected, estimator $\widehat{\sigma}$ based on LASSO produces estimates that are somewhat higher than the true value. In contrast, the estimator $\widehat{\sigma}$ based on Post-LASSO seems to perform very well in our experiments, giving estimates $\widehat{\sigma}$ that bunch closely near the true value σ.

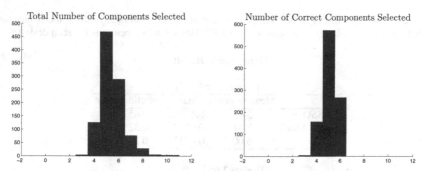

Fig. 3.5 The figure summarizes the covariate selection results for the design with $\sigma = 1$, when σ is estimated, based on 1000 Monte Carlo repetitions. The left panel plots the histogram for the number of covariates selected out of the possible 500 covariates. The right panel plots the histogram for the number of significant covariates selected; there are in total 6 significant covariates amongst 500 covariates. The sample size for each repetition was $n = 100$.

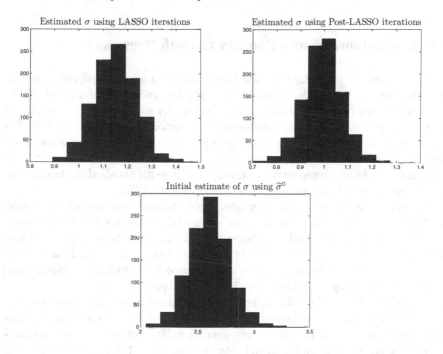

Fig. 3.6 The figure displays the distribution of the estimator $\widehat{\sigma}$ of σ based on (iterative) LASSO, (iterative) Post-LASSO, and the conservative initial estimator $\widehat{\sigma}^0 = \sqrt{\mathrm{Var}_n[y_i]}$. The plots summarize the estimation performance for the design with $\sigma = 1$, based on 1000 Monte Carlo repetitions.

Table 3.3 The table displays the average ℓ_0-norm of the estimators as well as mean bias and prediction error. We obtained the results using 1000 Monte Carlo repetitions for each design.

Monte Carlo Results

Design 1 ($\sigma^2 = 1$)			
	Mean ℓ_0-norm	Bias	Prediction Error
LASSO	5.41	0.4136	0.6572
Post-LASSO	5.41	0.0998	0.3298
Oracle	6.00	0.0122	0.2326

Design 2 ($\sigma^2 = 0.1$)			
	Mean ℓ_0-norm	Bias	Prediction Error
LASSO	6.3640	0.1395	0.2183
Post-LASSO	6.3640	0.0068	0.0893
Oracle	6.00	0.0039	0.0736

3.8 Application to Cross-Country Growth Regression

In this section we apply LASSO and Post-LASSO to an international economic growth example. We use the Barro and Lee [4] data consisting of a panel of 138 countries for the period of 1960 to 1985. We consider the national growth rates in GDP per capita as a dependent variable y for the periods 1965-75 and 1975-85. [2] In our analysis, we will consider a model with $p = 62$ covariates, which allows for a total of $n = 90$ complete observations. Our goal here is to select a subset of these covariates and briefly compare the resulting models to the standard models used in the empirical growth literature (Barro and Sala-i-Martin [5]).

Let us now turn to our empirical results. We performed covariate selection using LASSO, where we used our data-driven choice of penalty level λ in two ways. First we used an upper bound on σ being $\hat{\sigma}^0$ and decreased the penalty to estimate different models with $\lambda, \lambda/2, \lambda/3, \lambda/4$, and $\lambda/5$. Second, we applied the iterative procedure described in the previous section to define λ^{it} (which is computed based on $\hat{\sigma}^{it}$ obtained using the iterative Post-LASSO procedure).

The initial choice of the first approach led us to select no covariates, which is consistent with over-regularization since an upper bound for σ was used. We then proceeded to slowly decrease the penalty level in order to allow for some covariates to be selected. We present the model selection results in Table 3.5. With the first relaxation of the choice of λ, we select the black market exchange rate premium (characterizing trade openness) and a measure of political instability. With a second relaxation of the choice of λ we select an additional set of variables reported in the table. The iterative approach led to a model with only the black market exchange premium. We refer the reader to [4] and [5] for a complete definition and discussion of each of these variables.

---

[2] The growth rate in GDP over a period from t_1 to t_2 is commonly defined as $\log(GDP_{t_2}/GDP_{t_1})$.

We then proceeded to apply ordinary linear regression to the selected models and we also report the standard confidence intervals for these estimates. Table 3.4 shows these results. We find that in all models with additional selected covariates, the linear regression coefficients on the initial level of GDP is always negative and the standard confidence intervals do not include zero. We believe that these empirical findings firmly support the hypothesis of (conditional) convergence derived from the classical Solow-Swan-Ramsey growth model.[3] Finally, our findings also agree with and thus support the previous findings reported in Barro and Sala-i-Martin [5], which relied on ad-hoc reasoning for covariate selection.

Table 3.4 The table above displays the coefficient and a 90% confidence interval associated with each model selected by the corresponding penalty level. The selected models are displayed in Table 3.5.

Confidence Intervals after Model Selection
for the International Growth Regressions

Penalization Parameter $\lambda = 2.7870$	Real GDP per capita (log)	
	Coefficient	90% Confidence Interval
$\lambda'' = 2.3662$	−0.0112	[−0.0219, −0.0007]
$\lambda/2$	−0.0120	[−0.0225, −0.0015]
$\lambda/3$	−0.0153	[−0.0261, −0.0045]
$\lambda/4$	−0.0221	[−0.0346, −0.0097]
$\lambda/5$	−0.0370	[−0.0556, −0.0184]

Acknowledgements We would like to thank Denis Chetverikov and Brigham Fradsen for thorough proof-reading of several versions of this paper and their detailed comments that helped us considerably improve the paper. We also would like to thank Eric Gautier, Alexandre Tsybakov, and two anonymous referees for their comments that also helped us considerably improve the chapter. We would also like to thank the participants of seminars in Cowles Foundation Lecture at the Econometric Society Summer Meeting, Duke University, Harvard-MIT, and the Stats in the Chateau.

Appendix

3.9 Proofs

Proof (Theorem 3.1). Proceeding similarly to [8], by optimality of $\widehat{\beta}$ we have that

$$\widehat{Q}(\widehat{\beta}) - \widehat{Q}(\beta_0) \leqslant \frac{\lambda}{n}\|\beta_0\|_1 - \frac{\lambda}{n}\|\widehat{\beta}\|_1. \qquad (3.36)$$

---

[3] The inferential method used here is actually valid under certain conditions, despite the fact that the model has been selected; this is demonstrated in a work in progress.

Table 3.5 The models selected at various levels of penalty.

Model Selection Results for the International Growth Regressions

Penalization Parameter $\lambda = 2.7870$	Real GDP per capita (log) is included in all models Additional Selected Variables
λ	-
λ^{it}	Black Market Premium (log)
$\lambda/2$	Black Market Premium (log)
	Political Instability
$\lambda/3$	Black Market Premium (log)
	Political Instability
	Ratio of nominal government expenditure on defense to nominal GDP
	Ratio of import to GDP
$\lambda/4$	Black Market Premium (log)
	Political Instability
	Ratio of nominal government expenditure on defense to nominal GDP
$\lambda/5$	Black Market Premium (log)
	Political Instability
	Ratio of nominal government expenditure on defense to nominal GDP
	Ratio of import to GDP
	Exchange rate
	% of "secondary school complete" in male population
	Terms of trade shock
	Measure of tariff restriction
	Infant mortality rate
	Ratio of real government "consumption" net of defense and education
	Female gross enrollment ratio for higher education

To prove the result we make the use of the following relations: for $\delta = \widehat{\beta} - \beta_0$, if $\lambda \geqslant cn \|S\|_\infty$

$$\widehat{Q}(\widehat{\beta}) - \widehat{Q}(\beta_0) - \|\delta\|_{2,n}^2 = -2\mathbb{E}_n[\varepsilon_i x_i'\delta] - 2\mathbb{E}_n[r_i x_i'\delta] \tag{3.37}$$

$$\geqslant -\|S\|_\infty \|\delta\|_1 - 2c_s \|\delta\|_{2,n}$$

$$\geqslant -\frac{\lambda}{cn}(\|\delta_T\|_1 + \|\delta_{T^c}\|_1) - 2c_s \|\delta\|_{2,n}, \tag{3.38}$$

$$\|\beta_0\|_1 - \|\widehat{\beta}\|_1 = \|\beta_{0T}\|_1 - \|\widehat{\beta}_T\|_1 - \|\widehat{\beta}_{T^c}\|_1 \leqslant \|\delta_T\|_1 - \|\delta_{T^c}\|_1. \tag{3.39}$$

Thus, combining (3.36) with (3.37)–(3.39) implies that

$$-\frac{\lambda}{cn}(\|\delta_T\|_1 + \|\delta_{T^c}\|_1) + \|\delta\|_{2,n}^2 - 2c_s \|\delta\|_{2,n} \leqslant \frac{\lambda}{n}(\|\delta_T\|_1 - \|\delta_{T^c}\|_1). \tag{3.40}$$

If $\|\delta\|_{2,n}^2 - 2c_s \|\delta\|_{2,n} < 0$, then we have established the bound in the statement of the theorem. On the other hand, if $\|\delta\|_{2,n}^2 - 2c_s \|\delta\|_{2,n} \geqslant 0$ we get

$$\|\delta_{T^c}\|_1 \leqslant \frac{c+1}{c-1} \cdot \|\delta_T\|_1 = \overline{c} \|\delta_T\|_1, \tag{3.41}$$

and therefore δ satisfies the condition to invoke RE(c). From (3.40) and using RE(c), $\|\delta_T\|_1 \leqslant \sqrt{s}\|\delta\|_{2,n}/\kappa_{\overline{c}}$ we get

$$\|\delta\|_{2,n}^2 - 2c_s\|\delta\|_{2,n} \leqslant \left(1 + \frac{1}{c}\right)\frac{\lambda}{n}\|\delta_T\|_1 \leqslant \left(1 + \frac{1}{c}\right)\frac{\lambda\sqrt{s}}{n}\frac{\|\delta\|_{2,n}}{\kappa_{\overline{c}}}$$

which gives the result on the prediction norm.

Lemma 3.5 (Empirical pre-sparsity for LASSO). *In either the parametric model or the nonparametric model, let $\widehat{m} = |\widehat{T} \setminus T|$ and $\lambda \geqslant c \cdot n\|S\|_\infty$. We have*

$$\sqrt{\widehat{m}} \leqslant \sqrt{s}\sqrt{\phi(\widehat{m})}\, 2\overline{c}/\kappa_{\overline{c}} + 3(\overline{c}+1)\sqrt{\phi(\widehat{m})}\, nc_s/\lambda,$$

where $c_s = 0$ in the parametric model.

Proof. We have from the optimality conditions that

$$2\mathbb{E}_n[x_{ij}(y_i - x_i'\widehat{\beta})] = \text{sign}(\widehat{\beta}_j)\lambda/n \text{ for each } j \in \widehat{T} \setminus T.$$

Therefore we have for $R = (r_1, \ldots, r_n)'$, $X = [x_1, \ldots, x_n]'$, and $Y = (y_1, \ldots, y_n)'$

$$\begin{aligned}
\sqrt{\widehat{m}}\lambda &= 2\|(X'(Y - X\widehat{\beta}))_{\widehat{T}\setminus T}\| \\
&\leqslant 2\|(X'(Y - R - X\beta_0))_{\widehat{T}\setminus T}\| + 2\|(X'R)_{\widehat{T}\setminus T}\| + 2\|(X'X(\beta_0 - \widehat{\beta}))_{\widehat{T}\setminus T}\| \\
&\leqslant \sqrt{\widehat{m}} \cdot n\|S\|_\infty + 2n\sqrt{\phi(\widehat{m})}c_s + 2n\sqrt{\phi(\widehat{m})}\|\widehat{\beta} - \beta_0\|_{2,n},
\end{aligned}$$

where we used that

$$\begin{aligned}
\|(X'X(\beta_0 - \widehat{\beta}))_{\widehat{T}\setminus T}\| &\leqslant \sup_{\|v_{T^c}\|_0 \leqslant \widehat{m}, \|v\| \leqslant 1} |v'X'X(\beta_0 - \widehat{\beta})| \\
&\leqslant \sup_{\|v_{T^c}\|_0 \leqslant \widehat{m}, \|v\| \leqslant 1} \|v'X'\|\|X(\beta_0 - \widehat{\beta})\| \\
&= \sup_{\|v_{T^c}\|_0 \leqslant \widehat{m}, \|v\| \leqslant 1} \sqrt{|v'X'Xv|}\|X(\beta_0 - \widehat{\beta})\| \\
&= n\sqrt{\phi(\widehat{m})}\|\beta_0 - \widehat{\beta}\|_{2,n},
\end{aligned}$$

and similarly $\|(X'R)_{\widehat{T}\setminus T}\| \leqslant n\sqrt{\phi(\widehat{m})}c_s$.

Since $\lambda/c \geqslant n\|S\|_\infty$, and by Theorem 3.1, $\|\beta_0 - \widehat{\beta}\|_{2,n} \leqslant \left(1 + \frac{1}{c}\right)\frac{\lambda\sqrt{s}}{n\kappa_{\overline{c}}} + 2c_s$, we have

$$(1 - 1/c)\sqrt{\widehat{m}} \leqslant 2\sqrt{\phi(\widehat{m})}(1 + 1/c)\sqrt{s}/\kappa_{\overline{c}} + 6\sqrt{\phi(\widehat{m})}\, nc_s/\lambda.$$

The result follows by noting that $(1 - 1/c) = 2/(\overline{c}+1)$ by definition of $\overline{c}$.

Proof (Proof of Theorem 3.2). Since $\lambda \geqslant c \cdot n\|S\|_\infty$ by Lemma 3.5 we have

$$\sqrt{\widehat{m}} \leqslant \sqrt{\phi(\widehat{m})} \cdot 2\overline{c}\sqrt{s}/\kappa_{\overline{c}} + 3(\overline{c}+1)\sqrt{\phi(\widehat{m})} \cdot nc_s/\lambda,$$

which, by letting $L = \left(\frac{2\overline{c}}{\kappa_{\overline{c}}} + 3(\overline{c}+1)\frac{nc_s}{\lambda\sqrt{s}}\right)^2$, can be rewritten as

$$\widehat{m} \leqslant s \cdot \phi(\widehat{m})L. \tag{3.42}$$

Note that $\widehat{m} \leqslant n$ by optimality conditions. Consider any $M \in \mathcal{M}$, and suppose $\widehat{m} > M$. Therefore by Lemma 3.9 on sublinearity of sparse eigenvalues

$$\widehat{m} \leqslant s \cdot \left\lceil \frac{\widehat{m}}{M} \right\rceil \phi(M)L.$$

Thus, since $\lceil k \rceil < 2k$ for any $k \geqslant 1$ we have

$$M < s \cdot 2\phi(M)L$$

which violates the condition of $M \in \mathcal{M}$ and s. Therefore, we must have $\widehat{m} \leqslant M$.

In turn, applying (3.42) once more with $\widehat{m} \leqslant (M \wedge n)$ we obtain

$$\widehat{m} \leqslant s \cdot \phi(M \wedge n)L.$$

The result follows by minimizing the bound over $M \in \mathcal{M}$.

Proof (Lemma 3.3, part (1)). The result follows immediately from the assumptions.

Proof (Lemma 3.3, part (2)). Let $\widehat{m} = |\widehat{T} \setminus T| = \|\widehat{\beta}_{T^c}\|_0$. Then, note that $\|\delta\|_\infty \leqslant \|\delta\| \leqslant \|\delta\|_{2,n}/\kappa(\widehat{m})$. The result follows from Theorem 3.1.

Proof (Lemma 3.3, part (3)). Let $\delta := \widehat{\beta} - \beta_0$. Note that by the first order optimality conditions of $\widehat{\beta}$ and the assumption on λ

$$\|\mathbb{E}_n[x_i x_i' \delta]\|_\infty \leqslant \|\mathbb{E}_n[x_i(y_i - x_i'\widehat{\beta})]\|_\infty + \|S/2\|_\infty + \|\mathbb{E}_n[x_i r_i]\|_\infty$$
$$\leqslant \frac{\lambda}{2n} + \frac{\lambda}{2cn} + \min\left\{\frac{\sigma}{\sqrt{n}}, c_s\right\}$$

since $\|\mathbb{E}_n[x_i r_i]\|_\infty \leqslant \min\left\{\frac{\sigma}{\sqrt{n}}, c_s\right\}$ by Lemma 3.6 below.

Next let e_j denote the jth-canonical direction. Thus, for every $j = 1, \ldots, p$ we have

$$|\mathbb{E}_n[e_j' x_i x_i' \delta] - \delta_j| = |\mathbb{E}_n[e_j'(x_i x_i' - I)\delta]| \leqslant \max_{1 \leqslant j,k \leqslant p} |(\mathbb{E}_n[x_i x_i' - I])_{jk}| \, \|\delta\|_1$$
$$\leqslant \|\delta\|_1/[Us].$$

Then, combining the two bounds above and using the triangle inequality we have

$$\|\delta\|_\infty \leqslant \|\mathbb{E}_n[x_i x_i' \delta]\|_\infty + \|\mathbb{E}_n[x_i x_i' \delta] - \delta\|_\infty \leqslant \left(1 + \frac{1}{c}\right)\frac{\lambda}{2n} + \min\left\{\frac{\sigma}{\sqrt{n}}, c_s\right\} + \frac{\|\delta\|_1}{Us}.$$

The result follows by Lemma 3.7 to bound $\|\delta\|_1$ and the arguments in [8] and [17] to show that the bound on the correlations imply that for any $C > 0$

$$\kappa_C \geqslant \sqrt{1 - s(1 + 2C)\|\mathbb{E}_n[x_i x_i' - I]\|_\infty}$$

so that $\kappa_{\bar{c}} \geqslant \sqrt{1 - [(1 + 2\bar{c})/U]}$ and $\kappa_{2\bar{c}} \geqslant \sqrt{1 - [(1 + 4\bar{c})/U]}$ under this particular design.

Lemma 3.6. *Under condition ASM, we have that*

$$\|\mathbb{E}_n[x_i r_i]\|_\infty \leqslant \min\left\{\frac{\sigma}{\sqrt{n}}, c_s\right\}.$$

Proof. First note that for every $j = 1, \ldots, p$, we have $|\mathbb{E}_n[x_{ij} r_i]| \leqslant \sqrt{\mathbb{E}_n[x_{ij}^2]\mathbb{E}_n[r_i^2]} = c_s$.

Next, by definition of β_0 in (3.9), for $j \in T$ we have

$$\mathbb{E}_n[x_{ij}(f_i - x_i'\beta_0)] = \mathbb{E}_n[x_{ij} r_i] = 0$$

since β_0 is a minimizer over the support of β_0. For $j \in T^c$ we have that for any $t \in \mathbb{R}$

$$\mathbb{E}_n[(f_i - x_i'\beta_0)^2] + \sigma^2\frac{s}{n} \leqslant \mathbb{E}_n[(f_i - x_i'\beta_0 - tx_{ij})^2] + \sigma^2\frac{s+1}{n}.$$

Therefore, for any $t \in \mathbb{R}$ we have

$$-\sigma^2/n \leqslant \mathbb{E}_n[(f_i - x_i'\beta_0 - tx_{ij})^2] - \mathbb{E}_n[(f_i - x_i'\beta_0)^2] = -2t\mathbb{E}_n[x_{ij}(f_i - x_i'\beta_0)] + t^2\mathbb{E}_n[x_{ij}^2].$$

Taking the minimum over t in the right hand side at $t^* = \mathbb{E}_n[x_{ij}(f_i - x_i'\beta_0)]$ we obtain

$$-\sigma^2/n \leqslant -(\mathbb{E}_n[x_{ij}(f_i - x_i'\beta_0)])^2$$

or equivalently, $|\mathbb{E}_n[x_{ij}(f_i - x_i'\beta_0)]| \leqslant \sigma/\sqrt{n}$.

Lemma 3.7. *If $\lambda \geqslant cn\|S\|_\infty$, then for $\bar{c} = (c+1)/(c-1)$ we have*

$$\|\hat{\beta} - \beta_0\|_1 \leqslant \frac{(1 + 2\bar{c})\sqrt{s}}{\kappa_{2\bar{c}}}\left[\left(1 + \frac{1}{c}\right)\frac{\lambda\sqrt{s}}{n\kappa_{\bar{c}}} + 2c_s\right] + \left(1 + \frac{1}{2\bar{c}}\right)\frac{2c}{c-1}\frac{n}{\lambda}c_s^2,$$

where $c_s = 0$ in the parametric case.

Proof. First, assume $\|\delta_{T^c}\|_1 \leqslant 2\bar{c}\|\delta_T\|_1$. In this case, by definition of the restricted eigenvalue, we have

$$\|\delta\|_1 \leqslant (1 + 2\bar{c})\|\delta_T\|_1 \leqslant (1 + 2\bar{c})\sqrt{s}\|\delta\|_{2,n}/\kappa_{2\bar{c}}$$

and the result follows by applying the first bound to $\|\delta\|_{2,n}$ since $\bar{c} > 1$.

On the other hand, consider the case that $\|\delta_{T^c}\|_1 > 2\bar{c}\|\delta_T\|_1$ which would already imply $\|\delta\|_{2,n} \leqslant 2c_s$. Moreover, the relation (3.40) implies that

$$\begin{aligned}
\|\delta_{T^c}\|_1 &\leqslant \bar{c}\|\delta_T\|_1 + \frac{c}{c-1}\frac{n}{\lambda}\|\delta\|_{2,n}(2c_s - \|\delta\|_{2,n}) \\
&\leqslant \bar{c}\|\delta_T\|_1 + \frac{c}{c-1}\frac{n}{\lambda}c_s^2 \\
&\leqslant \tfrac{1}{2}\|\delta_{T^c}\|_1 + \frac{c}{c-1}\frac{n}{\lambda}c_s^2.
\end{aligned}$$

Thus,

$$\|\delta\|_1 \leqslant \left(1 + \frac{1}{2\bar{c}}\right)\|\delta_{T^c}\|_1 \leqslant \left(1 + \frac{1}{2\bar{c}}\right)\frac{2c}{c-1}\frac{n}{\lambda}c_s^2.$$

The result follows by adding the bounds on each case and invoking Theorem 3.1 to bound $\|\delta\|_{2,n}$.

Proof (Theorem 3.3). Let $\tilde{\delta} := \tilde{\beta} - \beta_0$. By definition of the Post-LASSO estimator, it follows that $\widehat{Q}(\tilde{\beta}) \leqslant \widehat{Q}(\widehat{\beta})$ and $\widehat{Q}(\tilde{\beta}) \leqslant \widehat{Q}(\beta_{0\widehat{T}})$. Thus,

$$\widehat{Q}(\tilde{\beta}) - \widehat{Q}(\beta_0) \leqslant \left(\widehat{Q}(\widehat{\beta}) - \widehat{Q}(\beta_0)\right) \wedge \left(\widehat{Q}(\beta_{0\widehat{T}}) - \widehat{Q}(\beta_0)\right) =: B_n \wedge C_n.$$

The least squares criterion function satisfies

$$
\begin{aligned}
|\widehat{Q}(\tilde{\beta}) - \widehat{Q}(\beta_0) - \|\tilde{\delta}\|_{2,n}^2| &\leqslant |S'\tilde{\delta}| + 2c_s\|\tilde{\delta}\|_{2,n} \\
&\leqslant |S_T'\tilde{\delta}| + |S_{T^c}'\tilde{\delta}| + 2c_s\|\tilde{\delta}\|_{2,n} \\
&\leqslant \|S_T\|\|\tilde{\delta}\| + \|S_{T^c}\|_\infty\|\tilde{\delta}_{T^c}\|_1 + 2c_s\|\tilde{\delta}\|_{2,n} \\
&\leqslant \|S_T\|\|\tilde{\delta}\| + \|S_{T^c}\|_\infty\sqrt{\widehat{m}}\|\tilde{\delta}\| + 2c_s\|\tilde{\delta}\|_{2,n} \\
&\leqslant \|S_T\|\frac{\|\tilde{\delta}\|_{2,n}}{\kappa(\widehat{m})} + \|S_{T^c}\|_\infty\sqrt{\widehat{m}}\frac{\|\tilde{\delta}\|_{2,n}}{\kappa(\widehat{m})} + 2c_s\|\tilde{\delta}\|_{2,n}.
\end{aligned}
$$

Next, note that for any $j \in \{1,\ldots,p\}$ we have $E[S_j^2] = 4\sigma^2/n$, so that $E[\|S_T\|^2] \leqslant 4\sigma^2 s/n$. Thus, by Chebyshev inequality, for any $\tilde{\gamma} > 0$, there is a constant $A_{\tilde{\gamma}}$ such that $\|S_T\| \leqslant A_{\tilde{\gamma}}\sigma\sqrt{s/n}$ with probability at least $1 - \tilde{\gamma}$. Moreover, using Lemma 3.8, $\|S_{T^c}\|_\infty \leqslant A_{\tilde{\gamma}}'2\sigma\sqrt{2\log p/n}$ with probability at least $1 - \tilde{\gamma}$ for some constant $A_{\tilde{\gamma}}'$. Define $A_{\gamma,n} := K_\gamma\sigma\sqrt{(s+\widehat{m}\log p)/n}$ so that $A_{\gamma,n} \geqslant \|S_T\| + \sqrt{\widehat{m}}\|S_{T^c}\|_\infty$ with probability at least $1 - \gamma$ for some constant $K_\gamma < \infty$ independent of n and p.

Combining these relations, with probability at least $1 - \gamma$ we have

$$\|\tilde{\delta}\|_{2,n}^2 - A_{\gamma,n}\|\tilde{\delta}\|_{2,n}/\kappa(\widehat{m}) - 2c_s\|\tilde{\delta}\|_{2,n} \leqslant B_n \wedge C_n,$$

solving which we obtain:

$$\|\tilde{\delta}\|_{2,n} \leqslant A_{\gamma,n}/\kappa(\widehat{m}) + 2c_s + \sqrt{(B_n)_+ \wedge (C_n)_+}. \tag{3.43}$$

Note that by the optimality of $\widehat{\beta}$ in the LASSO problem, and letting $\widehat{\delta} = \widehat{\beta} - \beta_0$,

$$B_n = \widehat{Q}(\widehat{\beta}) - \widehat{Q}(\beta_0) \leqslant \frac{\lambda}{n}(\|\beta_0\|_1 - \|\widehat{\beta}\|_1) \leqslant \frac{\lambda}{n}(\|\widehat{\delta}_T\|_1 - \|\widehat{\delta}_{T^c}\|_1). \tag{3.44}$$

If $\|\widehat{\delta}_{T^c}\|_1 > \bar{c}\|\widehat{\delta}_T\|_1$, we have $\widehat{Q}(\widehat{\beta}) - \widehat{Q}(\beta_0) \leqslant 0$ since $\bar{c} \geqslant 1$. Otherwise, if $\|\widehat{\delta}_{T^c}\|_1 \leqslant \bar{c}\|\widehat{\delta}_T\|_1$, by RE($\bar{c}$) we have

$$B_n := \widehat{Q}(\widehat{\beta}) - \widehat{Q}(\beta_0) \leqslant \frac{\lambda}{n}\|\widehat{\delta}_T\|_1 \leqslant \frac{\lambda}{n}\frac{\sqrt{s}\|\widehat{\delta}\|_{2,n}}{\kappa_{\bar{c}}}. \tag{3.45}$$

The choice of λ yields $\lambda \geqslant cn\|S\|_\infty$ with probability $1 - \alpha$. Thus, by applying Theorem 3.1, which requires $\lambda \geqslant cn\|S\|_\infty$, we can bound $\|\widehat{\delta}\|_{2,n}$.

Finally, with probability $1 - \alpha - \gamma$ we have that (3.43) and (3.45) with $\|\widehat{\delta}\|_{2,n} \leqslant (1+1/c)\lambda\sqrt{s}/n\kappa_{\bar{c}} + 2c_s$ hold, and the result follows since if $T \subseteq \widehat{T}$ we have $C_n = 0$ so that $B_n \wedge C_n \leqslant 1\{T \not\subseteq \widehat{T}\}B_n$.

Proof (Theorem 3.4). Consider the case of Post-LASSO; the proof for LASSO is similar. Consider the case with $k = 1$, i.e. when $\widehat{\sigma} = \widehat{\sigma}^k$ for $k = 1$. Then we have

$$\left| \frac{\widehat{Q}(\widetilde{\beta})}{\sigma^2} - \frac{\mathbb{E}_n[\varepsilon_i^2]}{\sigma^2} \right| \leqslant \frac{\|\widetilde{\beta} - \beta_0\|_{2,n}^2}{\sigma^2} + \frac{\|S\|_\infty\|\widetilde{\beta} - \beta_0\|_1}{\sigma^2} +$$

$$+ \frac{2c_s\|\widetilde{\beta} - \beta_0\|_{2,n}}{\sigma^2} + \frac{2c_s\sqrt{\mathbb{E}_n[\varepsilon_i^2]}}{\sigma^2} + \frac{c_s^2}{\sigma^2} = o_P(1).$$

since $\|\widetilde{\beta} - \beta_0\|_{2,n} \lesssim_P \sigma\sqrt{(s/n)\log p}$ by Corollary 3.3 and by assumption on $\widehat{\sigma}^0$, $\|S\|_\infty \lesssim_P \sigma\sqrt{(1/n)\log p}$ by Lemma 3.8, $\|\widetilde{\beta} - \beta_0\|_1 \leqslant \sqrt{\widehat{s}}\|\widetilde{\beta} - \beta\|_2 \lesssim_P \sqrt{\widehat{s}}\|\widetilde{\beta} - \beta\|_{2,n}$ by condition RSE, $\widehat{s} \lesssim_P s$ by Corollary 3.2 and $c_s \lesssim \sigma\sqrt{s/n}$ by condition ASM, and $s\log p/n \to 0$ by assumption, and $\frac{\mathbb{E}_n[\varepsilon_i^2]}{\sigma^2} - 1 \to_P 0$ by the Chebyshev inequality. Finally, $n/(n - \widehat{s}) = 1 + o_P(1)$ since $\widehat{s} \lesssim_P s$ by Corollary 3.2 and $s\log p/n \to 0$. The result for $2 \leqslant k \leqslant I - 1$ follows by induction.

3.10 Auxiliary Lemmas

Recall that $\|S/(2\sigma)\|_\infty = \max_{1 \leqslant j \leqslant p} |\mathbb{E}_n[x_{ij}g_i]|$, where g_i are i.i.d. $N(0,1)$, for $i = 1, ..., n$, conditional on $X = [x_1', ..., x_n']'$, and $\mathbb{E}_n[x_{ij}^2] = 1$ for each $j = 1, ..., p$, and note that $P(n\|S/(2\sigma)\|_\infty \geqslant \Lambda(1 - \alpha|X)|X) = \alpha$ by definition.

Lemma 3.8. *We have that for $t \geqslant 0$:*

$$P(n\|S/(2\sigma)\|_\infty \geqslant t\sqrt{n}|X) \leqslant 2p(1 - \Phi(t)) \leqslant 2p\frac{1}{t}\phi(t),$$

$$\Lambda(1 - \alpha|X) \leqslant \sqrt{n}\Phi^{-1}(1 - \alpha/2p) \leqslant \sqrt{2n\log(2p/\alpha)},$$

$$P(n\|S/(2\sigma)\|_\infty \geqslant \sqrt{2n\log(2p/\alpha)}|X) \leqslant \alpha.$$

Proof. To establish the first claim, note that $\sqrt{n}\|S/2\sigma\|_\infty = \max_{1 \leqslant j \leqslant p} |Z_j|$, where $Z_j = \sqrt{n}\mathbb{E}_n[x_{ij}g_i]$ are $N(0,1)$ by g_i i.i.d. $N(0,1)$ conditional on X and by $\mathbb{E}_n[x_{ij}^2] = 1$ for each $j = 1, ..., p$. Then the first claim follows by observing that for $z \geqslant 0$ by the union bound $P(\max_{1 \leqslant j \leqslant p} |Z_j| > z) \leqslant pP(|Z_j| > z) = 2p(1 - \Phi(z))$ and by $(1 - \Phi(z)) = \int_z^\infty \phi(u)du \leqslant \int_z^\infty (u/z)\phi(u)dz \leqslant (1/z)\phi(z)$. The second and third claim follow by noting that $2p(1 - \Phi(t')) = \alpha$ at $t' = \Phi^{-1}(1 - \alpha/2p)$, and $2p\frac{1}{t''}\phi(t'') = \alpha$ at $t'' \leqslant \sqrt{2\log(2p/\alpha)}$, so that, in view of the first claim, $\Lambda(1 - \alpha|X) \leqslant \sqrt{n}t' \leqslant \sqrt{n}t''$.

Lemma 3.9 (Sub-linearity of restricted sparse eigenvalues). *For any integer* $k \geqslant 0$ *and constant* $\ell \geqslant 1$ *we have* $\phi(\lceil \ell k \rceil) \leqslant \lceil \ell \rceil \phi(k)$.

Proof. Let $W := \mathbb{E}_n[x_i x_i']$ and $\bar{\alpha}$ be such that $\phi(\lceil \ell k \rceil) = \bar{\alpha}' W \bar{\alpha}$, $\|\bar{\alpha}\| = 1$. We can decompose the vector $\bar{\alpha}$ so that

$$\bar{\alpha} = \sum_{i=1}^{\lceil \ell \rceil} \alpha_i, \text{ with } \sum_{i=1}^{\lceil \ell \rceil} \|\alpha_{iT^c}\|_0 = \|\bar{\alpha}_{T^c}\|_0 \text{ and } \alpha_{iT} = \bar{\alpha}_T / \lceil \ell \rceil,$$

where we can choose α_i's such that $\|\alpha_{iT^c}\|_0 \leqslant k$ for each $i = 1, ..., \lceil \ell \rceil$, since $\lceil \ell \rceil k \geqslant \lceil \ell k \rceil$. Note that the vectors α_i's have no overlapping support outside T. Since W is positive semi-definite, $\alpha_i' W \alpha_i + \alpha_j' W \alpha_j \geqslant 2 |\alpha_i' W \alpha_j|$ for any pair (i, j). Therefore

$$\phi(\lceil \ell k \rceil) = \bar{\alpha}' W \bar{\alpha} = \sum_{i=1}^{\lceil \ell \rceil} \sum_{j=1}^{\lceil \ell \rceil} \alpha_i' W \alpha_j$$

$$\leqslant \sum_{i=1}^{\lceil \ell \rceil} \sum_{j=1}^{\lceil \ell \rceil} \frac{\alpha_i' W \alpha_i + \alpha_j' W \alpha_j}{2} = \lceil \ell \rceil \sum_{i=1}^{\lceil \ell \rceil} \alpha_i' W \alpha_i$$

$$\leqslant \lceil \ell \rceil \sum_{i=1}^{\lceil \ell \rceil} \|\alpha_i\|^2 \phi(\|\alpha_{iT^c}\|_0) \leqslant \lceil \ell \rceil \max_{i=1,...,\lceil \ell \rceil} \phi(\|\alpha_{iT^c}\|_0) \leqslant \lceil \ell \rceil \phi(k),$$

where we used that

$$\sum_{i=1}^{\lceil \ell \rceil} \|\alpha_i\|^2 = \sum_{i=1}^{\lceil \ell \rceil} (\|\alpha_{iT}\|^2 + \|\alpha_{iT^c}\|^2) = \frac{\|\bar{\alpha}_T\|^2}{\lceil \ell \rceil} + \sum_{i=1}^{\lceil \ell \rceil} \|\alpha_{iT^c}\|^2 \leqslant \|\bar{\alpha}\|^2 = 1.$$

Lemma 3.10. *Let* $\bar{c} = (c+1)/(c-1)$ *we have for any integer* $m > 0$

$$\kappa_{\bar{c}} \geqslant \kappa(m) \left(1 - \mu(m) \bar{c} \sqrt{\frac{s}{m}}\right).$$

Proof. We follow the proof in [8]. Pick an arbitrary vector δ such that $\|\delta_{T^c}\|_1 \leqslant \bar{c}\|\delta_T\|_1$. Let T^1 denote the m largest components of δ_{T^c}. Moreover, let $T^c = \cup_{k=1}^K T^k$ where $K = \lceil (p-s)/m \rceil$, $|T^k| \leqslant m$ and T^k corresponds to the m largest components of δ outside $T \cup (\cup_{d=1}^{k-1} T^d)$.

We have

$$\|\delta\|_{2,n} \geqslant \|\delta_{T \cup T^1}\|_{2,n} - \|\delta_{(T \cup T^1)^c}\|_{2,n} \geqslant \kappa(m) \|\delta_{T \cup T^1}\| - \sum_{k=2}^K \|\delta_{T^k}\|_{2,n}$$

$$\geqslant \kappa(m) \|\delta_{T \cup T^1}\| - \sqrt{\phi(m)} \sum_{k=2}^K \|\delta_{T^k}\|.$$

Next note that

$$\|\delta_{T^{k+1}}\| \leqslant \|\delta_{T^k}\|_1 / \sqrt{m}.$$

Indeed, consider the problem $\max\{\|v\|/\|u\|_1 : v, u \in \mathbb{R}^m, \max_i |v_i| \leqslant \min_i |u_i|\}$. Given a v and u we can always increase the objective function by using $\tilde{v} = \max_i |v_i|$ $(1,\ldots,1)'$ and $\tilde{u}' = \min_i |u_i|(1,\ldots,1)'$ instead. Thus, the maximum is achieved at $v^* = u^* = (1,\ldots,1)'$, yielding $1/\sqrt{m}$.

Thus, by $\|\delta_{T^c}\|_1 \leqslant \bar{c}\|\delta_T\|_1$ and $|T| = s$

$$\sum_{k=2}^{K} \|\delta_{T^k}\| \leqslant \sum_{k=1}^{K-1} \frac{\|\delta_{T^k}\|_1}{\sqrt{m}} \leqslant \frac{\|\delta_{T^c}\|_1}{\sqrt{m}} \leqslant \bar{c}\|\delta_T\|\sqrt{\frac{s}{m}} \leqslant \bar{c}\|\delta_{T \cup T^1}\|\sqrt{\frac{s}{m}}.$$

Therefore, combining these relations with $\|\delta_{T \cup T^1}\| \geqslant \|\delta_T\| \geqslant \|\delta_T\|_1/\sqrt{s}$ we have

$$\|\delta\|_{2,n} \geqslant \frac{\|\delta_T\|_1}{\sqrt{s}} \kappa(m) \left(1 - \mu(m)\bar{c}\sqrt{s/m}\right)$$

which leads to

$$\frac{\sqrt{s}\|\delta\|_{2,n}}{\|\delta_T\|_1} \geqslant \kappa(m) \left(1 - \mu(m)\bar{c}\sqrt{s/m}\right).$$

References

1. Akaike, H.: A new look at the statistical model identification. IEEE Trans. Automatic Control AC-19, 716–723 (1974)
2. Angrist, J., Chernozhukov, V., Fernández-Val, I.: Quantile regression under misspecification, with an application to the U.S. wage structure. Econometrica 74(2), 539–563 (2006)
3. Angrist, J.D., Krueger, A.B.: Does compulsory school attendance affect schooling and earnings? Quart. J. Econom. 106(4), 979–1014 (1991)
4. Barro, R.J., Lee, J.W.: Data set for a panel of 138 countries. Tech. rep., NBER (1994). URL http://www.nber.org/pub/barro.lee/
5. Barro, R.J., Sala–i–Martin, X.: Economic Growth. McGraw-Hill, New York (1995)
6. Belloni, A., Chen, D., Chernozhukov, V., Hansen, C.: Sparse models and methods for optimal instruments with an application to eminent domain (2010). URL http://arxiv.org/abs/1010.4345
7. Belloni, A., Chernozhukov, V.: Post-ℓ_1-penalized estimators in high-dimensional linear regression models (2009). URL http://arxiv.org/abs/1001.0188
8. Bickel, P.J., Ritov, Y., Tsybakov, A.B.: Simultaneous analysis of LASSO and Dantzig selector. Ann. Statist. 37(4), 1705–1732 (2009)
9. Candès, E.J., Plan, Y.: Near-ideal model selection by ℓ_1 minimization. Ann. Statist. 37(5A), 2145–2177 (2009)
10. Candès, E.J., Tao, T.: The Dantzig selector: statistical estimation when p is much larger than n. Ann. Statist. 35(6), 2313–2351 (2007)
11. Donoho, D.L., Johnstone, J.M.: Ideal spatial adaptation by wavelet shrinkage. Biometrika 81(3), 425–455 (1994)
12. Ge, D., Jiang, X., Ye, Y.: A note on complexity of L_p minimization (2010). Stanford Working Paper
13. van de Geer, S.: High-dimensional generalized linear models and the Lasso. Ann. Statist. 36(2), 614–645 (2008)
14. Hansen, C., Hausman, J., Newey, W.K.: Estimation with many instrumental variables. J. Bus. Econom. Statist. 26, 398–422 (2008)

15. Koltchinskii, V.: Sparsity in penalized empirical risk minimization. Ann. Inst. Henri Poincaré Probab. Stat. **45**(1), 7–57 (2009)
16. Levine, R., Renelt, D.: A sensitivity analysis of cross-country growth regressions. The American Economic Review **82**(4), 942–963 (1992)
17. Lounici, K.: Sup-norm convergence rate and sign concentration property of Lasso and Dantzig estimators. Electron. J. Statist. **2**, 90–102 (2008)
18. Meinshausen, N., Yu, B.: Lasso-type recovery of sparse representations for high-dimensional data. Ann. Statist. **37**(1), 2246–2270 (2009)
19. Natarajan, B.K.: Sparse approximate solutions to linear systems. SIAM J. Comput. **24**, 227–234 (1995)
20. Rigollet, P., Tsybakov, A.B.: Exponential screening and optimal rates of sparse estimation. Ann. Statist. (2010). To appear
21. Sala–i–Martin, X.: I just ran two million regressions. Amer. Econ. Rev. **87**(2), 178–183 (1997)
22. Schwarz, G.: Estimating the dimension of a model. Ann. Statist. **6**, 461–464 (1978)
23. Tibshirani, R.: Regression shrinkage and selection via the Lasso. J. R. Stat. Soc. Ser. B Stat. Methodol. **58**, 267–288 (1996)
24. Wainwright, M.: Sharp thresholds for noisy and high-dimensional recovery of sparsity using ℓ_1-constrained quadratic programming (Lasso). IEEE Trans. Inform. Theory **55**, 2183–2202 (2009)
25. Zhang, C.H., Huang, J.: The sparsity and bias of the Lasso selection in high-dimensional linear regression. Ann. Statist. **36**(4), 1567–1594 (2008)
26. Zhao, P., Yu, B.: On model selection consistency of Lasso. J. Mach. Learn. Res. **7**, 2541–2567 (2006)

Part IV
Invited Contributions on
High-Dimensional Estimation

Chapter 4
Model Selection in Gaussian Regression for High-Dimensional Data

Felix Abramovich and Vadim Grinshtein

Abstract We consider model selection in Gaussian regression, where the number of predictors might be even larger than the number of observations. The proposed procedure is based on penalized least square criteria with a complexity penalty on a model size. We discuss asymptotic properties of the resulting estimators corresponding to linear and so-called $2k\ln(p/k)$-type nonlinear penalties for nearly-orthogonal and multicollinear designs. We show that any linear penalty cannot be simultaneously adapted to both sparse and dense setups, while $2k\ln(p/k)$-type penalties achieve the wide adaptivity range. We also present Bayesian perspective on the procedure that provides an additional insight and can be used as a tool for obtaining a wide class of penalized estimators associated with various complexity penalties.

4.1 Introduction

Modern statistics encounters new challenges, where the problems have exploded both in size and complexity. Analysis of complex high-dimensional data sets of very large sizes requires a new look on traditional statistical methods.

Consider the standard Gaussian linear regression setup

$$\mathbf{y} = X\beta + \varepsilon, \tag{4.1}$$

where $\mathbf{y} \in \mathbb{R}^n$ is a vector of the observed response variable Y, $X_{n \times p}$ is the design matrix of p explanatory variables (predictors) $X_1, ..., X_p$, $\beta \in \mathbb{R}^p$ is a vector of un-

Felix Abramovich
Tel Aviv University, Department of Statistics & Operations Research, Ramat Aviv, Tel Aviv 69978, Israel, e-mail: felix@post.tau.ac.il

Vadim Grinshtein
The Open University of Israel, Department of Mathematics, P.O.Box 808, Raanana 43107, Israel, e-mail: vadimg@openu.ac.il

known regression coefficients, $\varepsilon \sim N(\mathbf{0}, \sigma^2 I_n)$ and the noise variance σ^2 is assumed to be known.

The number of predictors p might be very large relatively even to the amount of available data n that causes a severe "curse of dimensionality" problem. However, it is usually believed that only a small fraction of them has a truly relevant impact on the response. Thus, the problem of model (or variable) selection for reduction dimensionality in (4.1) becomes of fundamental importance. Its main goal is to select the "best", parsimonious subset of predictors (model) among $X_1, ..., X_p$. For a selected model M, the corresponding coefficients β_M are then typically estimated by least squares. The definition of the "best" subset however depends on the particular aim at hand. One should distinguish, for example, between estimating regression coefficients β, estimating the mean vector $X\beta$, identifying non-zero coefficients and predicting future observations. Different aims may lead to different optimal model selection procedures especially for the "p larger than n" setup. In this paper we focus on estimating the mean vector $X\beta$ and the goodness of a given model M is measured by the quadratic risk $E||X\hat{\beta}_M - X\beta||^2 = ||X\beta_M - X\beta||^2 + \sigma^2|M|$, where $X\beta_M$ is the projection of $X\beta$ on the span of M and $\hat{\beta}_M$ is the least square estimate of β_M. The bias term represents the approximation error of the projection, while the variance term is the price for estimating the projection coefficients β_M by $\hat{\beta}_M$ and is proportional to the model size. The "best" model then is the one with the minimal quadratic risk. Note that the true underlying model in (4.1) is not necessarily the best in this sense since sometimes it is possible to reduce its risk by excluding predictors with small (but still nonzero!) coefficients.

Since the above criterion involves the unknown β, the corresponding ideal minimal risk can be rather used as a benchmark for any available model selection procedure. Typical model selection criterion is based on the *empirical* quadratic risk $||\mathbf{y} - X\hat{\beta}_M||^2$, which is essentially the least squares. The empirical risk obviously decreases as the model size grows and to avoid overfitting, it is penalized by a complexity penalty $Pen(|M|)$ that increases with $|M|$. This leads to the *penalized* least square criterion of the form

$$||\mathbf{y} - X\hat{\beta}_M||^2 + Pen(|M|) \to \min_M \qquad (4.2)$$

The properties of the resulting estimator depends on the proper choice of the complexity penalty $Pen(\cdot)$ in (4.2). A large amount of works has studied various types of penalties. The most commonly used choice is a *linear* type penalty of the form $Pen(k) = 2\sigma^2 \lambda k$ for some fixed $\lambda > 0$. The most known examples motivated by a wide variety of approaches include C_p (Mallows, 1973) and AIC (Akaike, 1974) for $\lambda = 1$, BIC (Schwarz, 1978) for $\lambda = (\ln n)/2$ and RIC (Foster & George, 1994) for $\lambda = \ln p$. On the other hand, a series of recent works suggested the so-called $2k \ln(p/k)$-type *nonlinear* complexity penalties of the form

$$Pen(k) = 2\sigma^2 ck(\ln(p/k) + \zeta_{p,k}), \qquad (4.3)$$

where $c > 1$ and $\zeta_{p,k}$ is some "negligible" term (see, e.g., Birgé & Massart, 2001, 2007; Johnstone, 2002; Abramovich *et al.*, 2006; Bunea, Tsybakov & Wegkamp, 2007; Abramovich & Grinshtein, 2010).

In this paper we discuss the asymptotic properties of linear and $2k\ln(p/k)$-type penalized estimators (4.2) as both the sample size n and the number of predictors p increase. We distinguish between two different types of the design: *nearly-orthogonal*, where there is no strong collinearity between predictors, and *multi-collinear*, that usually appears when $p \gg n$. Interesting, that the minimax rates for estimating the mean vector for multicollinear design are faster than those for nearly-orthogonal by a certain factor depending on the design properties. Such a phenomenon can be explained by a possibility of exploiting strong correlations between predictors to reduce the model size without paying much extra price in the bias.

We show that even for nearly-orthogonal design any linear penalty cannot be simultaneously optimal (in the minimax sense) for both sparse and dense cases. On the contrary, the $2k\ln(p/k)$-types penalties achieve the widest possible adaptivity range. Moreover, under some additional assumptions on the design and regression coefficients vector, they remain asymptotically optimal for the multicollinear design as well.

We also describe a Bayesian interpretation of penalized estimators (4.2) developed in Abramovich & Grinshtein (2010) for a general case and for the considered two types of penalties in particular. Bayesian approach provides an additional insight in these estimators and can be also used as a tool for obtaining a wide class of penalized estimators with various complexity penalties.

The paper is organized as follows. The notations, definitions and some preliminary results are given in Section 4.2, where, in particular, we present the (nonasymptotic) minimax lower bound for the risk of estimating the means vector $X\beta$ in (4.1). The asymptotic minimax properties of penalized estimators (4.2) for nearly-orthogonal and multicollinear designs are investigated respectively in Sections 4.3 and 4.4. Section 4.5 presents a Bayesian perspective on (4.2). Some concluding remarks are given in Section 4.6.

4.2 Preamble

Consider the general linear regression setup (4.1), where the number of possible predictors p may be even larger then the number of observations n. Let $r = rank(X)(\leqslant \min(p,n))$ and assume that any r columns of X are linearly independent. For the "standard" linear regression setup, where all p predictors are linearly independent and there are at least p linearly independent design points, $r = p$.

Any model M is uniquely defined by the $p \times p$ diagonal indicator matrix $D_M = diag(\mathbf{d}_M)$, where $d_{jM} = \mathbb{I}\{X_j \in M\}$ and, therefore, $|M| = tr(D_M)$. For a given M, the least square estimate of its coefficients is $\hat{\beta}_M = (D_M X'X D_M)^+ D_M X'\mathbf{y}$, where "+" denotes the generalized inverse matrix.

For a fixed p_0, define the sets of models $\mathcal{M}_{p_0}$ that have at most p_0 predictors, that is, $\mathcal{M}_{p_0} = \{M : |M| \leqslant p_0\}$. Obviously, if a true model in (4.1) belongs to $\mathcal{M}_{p_0}$, then $||\beta||_0 \leqslant p_0$, where the l_0 quasi-norm of the coefficients vector β is defined as the number of its nonzero entries. We consider $p_0 \leqslant r$ since otherwise, there necessarily exists another vector β^* such that $||\beta^*||_0 \leqslant r$ and $X\beta = X\beta^*$.

Within the minimax framework, the performance of a penalized estimator $X\hat{\beta}_{\hat{M}}$ of the unknown mean vector $X\beta$ in (4.1) corresponding to the selected model $\hat{M}$ with respect to (4.2) over $\mathcal{M}_{p_0}$ is measured by its worst-case quadratic risk $\sup_{\beta:||\beta||_0 \leqslant p_0} E||X\hat{\beta}_{\hat{M}} - X\beta||^2$. It is then compared to the minimax risk – the best attainable worst-case risk among all possible estimators,

$$R(\mathcal{M}_{p_0}) = \inf_{\hat{y}} \sup_{\beta:||\beta||_0 \leqslant p_0} E||\hat{y} - X\beta||^2.$$

We present first the following result of Abramovich & Grinshtein (2010) for the lower bound for the minimax risk $R(\mathcal{M}_{p_0})$.

For any given $k = 1, ..., r$, let $\phi_{min}[k]$ and $\phi_{max}[k]$ be the k-sparse minimal and maximal eigenvalues of the design defined as

$$\phi_{min}[k] = \min_{\beta:1 \leqslant ||\beta||_0 \leqslant k} \frac{||X\beta||^2}{||\beta||^2},$$

$$\phi_{max}[k] = \max_{\beta:1 \leqslant ||\beta||_0 \leqslant k} \frac{||X\beta||^2}{||\beta||^2}$$

In fact, $\phi_{min}[k]$ and $\phi_{max}[k]$ are respectively the minimal and maximal eigenvalues of all $k \times k$ submatrices of the matrix $X'X$ generated by any k columns of X. Let $\tau[k] = \phi_{min}[k]/\phi_{max}[k]$, $k = 1, ..., r$. By the definition, $\tau[k]$ is a non-increasing function of k. Obviously, $\tau[k] \leqslant 1$ and for the orthogonal design the equality holds for all k.

Theorem 4.1. *(Abramovich & Grinshtein, 2010). Consider the model (4.1) and let $1 \leqslant p_0 \leqslant r$. There exists a universal constant $C > 0$ such that*

$$R(\mathcal{M}_{p_0}) \geqslant \begin{cases} C\sigma^2 \tau[2p_0] \, p_0(\ln(p/p_0) + 1) \,, & 1 \leqslant p_0 \leqslant r/2 \\ C\sigma^2 \tau[p_0] \, r & , \, r/2 \leqslant p_0 \leqslant r \end{cases} \qquad (4.4)$$

Note that the minimax lower bound (4.4) depends on a design matrix X only through its sparse eigenvalues ratios. Computationally simpler but less accurate minimax lower bound can be obtained by replacing $\tau[2p_0]$ and $\tau[p_0]$ in (4.4) by $\tau[r]$, that for the case $r = p \leqslant n$ is just the ratio of the minimal and maximal eigenvalues of $X'X$.

Consider now the asymptotics as the sample size n increases. We allow $p = p_n$ to increase with n as well in such a way that r tends to infinity and look for a projection of the unknown mean vector on an expanding span of predictors. In the "classical" regression setup, $p_n = o(n)$, while in the "modern" one, p_n may be larger than n or even $p_n \gg n$.

In such asymptotic setting one should essentially consider a *sequence* of design matrices X_{n,p_n}, where $r_n \to \infty$. For simplicity of exposition, in what follows the index n is omitted and X_{n,p_n} will be denoted by X_p emphasizing the dependence on the number of predictors p when r tend to infinity. Similarly, we consider now sequences of corresponding coefficients vectors β_p. In these notations, the original model (4.1) is transformed into a sequence of models

$$\mathbf{y} = X_p \beta_p + \varepsilon, \tag{4.5}$$

where $rank(X) = r$ and any r columns of X are linearly independent (hence, $\tau_p[r] > 0$), $\varepsilon \sim N(\mathbf{0}, \sigma^2 I_n)$ and the noise variance σ^2 does not depend on n and p.

The minimax lower bound (4.4) indicates that depending on the asymptotic behavior of the sparse eigenvalues ratios, one should distinguish between nearly-orthogonal and multicollinear designs:

Definition 4.1. Consider the sequence of design matrices X_p. The design is called nearly-orthogonal if the corresponding sequence of sparse eigenvalues ratios $\tau_p[r]$ is bounded away from zero by some constant $c > 0$. Otherwise, the design is called multicollinear.

Nearly-orthogonality assumption essentially means that there is no collinearity in the design in the sense that there are no "too strong" linear relationships within any set of r columns of X_p. It is intuitively clear that it can happen only when p is not "too large" relative to r (and hence to n), while for the $p_n \gg n$ setup, multicollinearity between predictors is inherent. Indeed, Abramovich & Grinshtein (2010) showed that for nearly-orthogonal design necessarily $p = O(r)$ and, thefore, $p = O(n)$.

In what follows we consider separately the two types of the design and investigate the asymptotic optimality (in the minimax sense) of linear and $2k\ln(p/k)$-type penalties.

4.3 Nearly-Orthogonal Design

From the definition of nearly-orthogonal design it follows that there exists a constant $c > 0$ such that $c \leqslant \tau_p[r] \leqslant ... \leqslant \tau_p[1] = 1$. In addition, as we have mentioned in the previous Section 4.2, for this type of design $p = O(r)$ and, therefore, the minimax lower bound (4.4) over $\mathcal{M}_{p_0}$ in this case is essentially of the order $p_0(\ln(p/p_0) + 1)$ for all $p_0 = 1, ..., r$.

We start from linear penalties, where $Pen(k) = 2\sigma^2 \lambda_p k$. Foster & George (1994) and Birgé & Massart (2001, Section 5.2) showed that the best possible risk of corresponding penalized estimators over $\mathcal{M}_{p_0}$ is of the order $\sigma^2 p_0 \ln p$ achieved for $\lambda_p = (1 + \delta)\ln p$, $\delta > 0$ corresponding to the RIC criterion. This risk is of the same order as $p_0(\ln(p/p_0) + 1)$ in the minimax lower bound (4.4) when $p_0 = O(r^\alpha)$ for some $0 < \alpha < 1$ (sparse cases), but higher than the latter for the dense cases, where $p_0 \sim r$. On the other hand, it is the AIC estimator ($\lambda_p = 1$) with the risk of the order $\sigma^2 p$, that is asymptotically similar to (4.4) for dense but much higher for

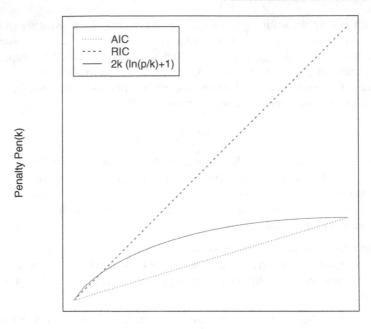

Fig. 4.1 Various penalties: AIC (dotted line), RIC (dashed line) and $2k(\ln(p/k)+1)$ (solid line).

sparse cases. In other words, no penalized estimator (4.2) with a linear penalty can be simultaneously rate-optimal for both sparse and dense cases. Note that a linear penalty $Pen(k) = 2\sigma^2\lambda_p k$ yields the *constant* per predictor price $2\sigma^2\lambda_p$ that cannot be adapted to both cases.

On the other hand, the nonlinear penalties of the $2k\ln(p/k)$-type imply *different* per predictor price: higher for small models but decreasing as the model size grows. In fact, such type of penalty behaves like RIC for sparse and AIC for dense cases (see Figure 4.1). As we shall show, it allows the corresponding estimators to achieve the widest adaptivity range.

Consider a general $2k\ln(p/k)$-type penalty (4.3), where $c > 1$. From the results of Birgé & Massart (2001, 2007) and Abramovich & Grinshtein (2010) for the corresponding penalized estimators (4.2) it follows that for any $1 \leqslant p_0 \leqslant r$,

$$\sup_{\beta:\|\beta\|_0 \leqslant p_0} E\|X\hat{\beta}_{\hat{M}} - X\beta\|^2 \leqslant C\sigma^2 p_0(\ln(p/p_0)+1) \tag{4.6}$$

for some $C > 0$. Comparing the risk upper bound (4.6) with the minimax lower bound implies the following Corollary:

Corollary 4.1. *Let the design be nearly-orthogonal. Consider the penalized least square estimation (4.2) with a $2k\ln(p/k)$-type complexity penalty (4.3), where $c > 1$. Then, as $r \to \infty$, the corresponding penalized estimator attains the minimax convergence rates simultaneously over all $\mathcal{M}_{p_0}$, $p_0 = 1, ..., r$.*

Furthermore, for sparse cases Birgé & Massart (2007) showed that for $c < 1$ in (4.3), the risk of the corresponding penalized estimator is much larger than that in (4.6). The value $c = 1$ for the $2k\ln(p/k)$-type penalty (4.3) is, therefore, a borderline. For the *orthogonal* design and various sparse settings, Abramovich *et al.* (2006) and Wu & Zhou (2009) proved that $c = 1$ yields even *sharp* (with an exact constant) asymptotic minimaxity. However, to the best of our knowledge, it is not clear what happens for the choice $c = 1$ in a general case.

Finally, note that for the nearly-orthogonal design, $||X_p\hat{\beta}_{p\hat{M}} - X_p\beta_p|| \asymp ||\hat{\beta}_{p\hat{M}} - \beta_p||$, where "$\asymp$" means that their ratio is bounded from below and above. Therefore, all the results of this section for estimating the mean vector $X_p\beta_p$ in (4.5) can be straightforwardly applied for estimating the regression coefficients β_p. This equivalence, however, does not hold for the multicollinear design considered below.

4.4 Multicollinear Design

Recall that nearly-orthogonality assumption necessarily implies $p = O(n)$. Thus, it may be reasonable in the "classical" setup, where p is not too large relatively to n but is questionable for the analysis of high-dimensional data, where $p \gg n$. In this section we investigate the performance of $2k\ln(p/k)$-type penalties for the multicollinear design.

When nearly-orthogonality does not hold, the sparse eigenvalues ratios in (4.4) may tend to zero as p increases and, thus, decrease the minimax lower bound rate relatively to the nearly-orthogonal design. In this case there is a gap between the rates in the lower and upper bounds (4.4) and (4.6). Intuitively, strong correlations between predictors can be exploited to diminish the size of a model (hence, to decrease the variance) without paying much extra price in the bias, and, therefore, to reduce the overall risk. It turns out that under certain additional assumptions on the design and the regression coefficients vector in (4.5) given below, the upper risk bound (4.6) of the $2k\ln(p/k)$-type estimator can be indeed reduced to the minimax lower bound rate (4.4). Under these conditions, $2k\ln(p/k)$-type penalized estimator, therefore, remains asymptotically rate-optimal even for the multicollinear design.

For simplicity of exposition we consider $p_0 \leqslant r/2$ although the results for $r/2 \leqslant p_0 \leqslant r$ can be obtained in a similar way with necessary changes. In particular, for the latter case one should slightly modify the $2k\ln(p/k)$-type penalty for $k = r$ to be of the form $Pen(r) \sim 2\sigma^2 cr$ for some $c > 0$ (Abramovich & Grinshtein, 2010). Note that for the nearly-orthogonal design, where $p = O(r)$, $2k\ln(p/k)$-type penalties automatically imply this condition on $Pen(r)$.

For any model M of size $k \leqslant r/2$ let X_M be the $n \times k$ submatrix of X_p containing the corresponding k columns of X_p. Consider the matrix $(X_M'X_M)^{-1}$ and the maximum of minimal eigenvalues $\phi_{min}[(X'_M X_M)^{-1}]_{M'}$ of all its symmetric $\lfloor k(1 - \tau_p[2k]) \rfloor \times \lfloor k(1 - \tau_p[2k]) \rfloor$ submatrices corresponding to various submodels $M' \subset M$ of size $\lfloor k(1 - \tau_p[2k]) \rfloor$. Define $\tilde{\phi}_p[k] = \min_M \phi_{min}[(X'_M X_M)^{-1}]_{M'}$, that is,

$$\tilde{\phi}_p[k] = \min_{M:|M|=k} \quad \max_{\substack{M' \subset M \\ |M'| = \lfloor k(1 - w_p[k]) \rfloor}} \quad \phi_{min}[(X'_M X_M)^{-1}]_{M'}$$

It can be shown (Abramovich & Grinshtein, 2010) that $\tilde{\phi}_p^{-1}[k]$ measures an error of approximating mean vectors $X_p \beta_p$, where $||\beta_p||_0 = k$, by their projections on lower dimensional subspans of predictors. The stronger is multicollinearity, the better is the approximation and the larger is $\tilde{\phi}_p[k]$.

The following theorem is a consequence of Theorem 5 of Abramovich & Grinshtein (2010):

Theorem 4.2. *Let $\tau_p[r] \to 0$ as $r \to \infty$ (multicollinear design). Assume the following additional assumptions on the design matrix X_p and the (unknown) vector of coefficients β_p in (4.5):*

(D)   for all p there exist $1 \leqslant \kappa_{p1} \leqslant \kappa_{p2} \leqslant r/2$ such that

1. $\tilde{c}_1 \leqslant \tau_p[2k] \cdot k \leqslant k - 1$, $k = \kappa_{p1}, ..., \kappa_{p2}$
2. $\tau_p[2\kappa_{p2}] \geqslant (\kappa_{p2}/(pe))^{\tilde{c}_2}$
3. $\phi_{p,min}[2k] \cdot \tilde{\phi}_p[k] \geqslant \tilde{c}_3$, $k = \kappa_{p1}, ..., \kappa_{p2}$

(B) $||\beta_p||_\infty^2 \leqslant \tilde{c}_4 \tau_p[2p_0] \cdot \tilde{\phi}_p[p_0] \cdot (\ln(p/p_0) + 1)$, where $p_0 = ||\beta_p||_0$

for some positive constants $\tilde{c}_1$, $\tilde{c}_2$, $\tilde{c}_3$ and $\tilde{c}_4$.

Then, the penalized least square estimator (4.2) with a $2k\ln(p/k)$-type complexity penalty (4.3), where $c > 1$, is asymptotically simultaneously minimax (up to a constant multiplier) over all $\mathcal{M}_{p_0}$, $\kappa_{p1} \leqslant p_0 \leqslant \kappa_{p2}$.

Generally, Assumptions (D.1, D.2) and Assumption (B) allow one to reduce the upper bound (4.6) for the risk of the $2k\ln(p/k)$-type estimator by the factor $\tau_p[2p_0]$, while Assumption (D.3) is required to guarantee that the additional constraint on β_p in Assumption (B) does not affect the lower bound (4.4). We have mentioned that multicollinearity typically arises when $p \gg n$. One can easily verify that for $n = O(p^\alpha)$, $0 \leqslant \alpha < 1$, Assumption (D.2) always follows from Assumption (D.1) and, therefore, can be omitted in this case.

4.5 Bayesian Perspective

In this section we discuss the Bayesian approach to model selection in the Gaussian regression model (4.1) proposed by Abramovich & Grinshtein (2010). Bayesian framework naturally interpretates the penalized least square estimation (4.2) by treating the penalty term as proportional to the logarithm of a prior distribution on the model size. Minimization of (4.2) corresponds then to the maximum *a posteriori* (MAP) rule. Choosing different types of a prior, the resulting Bayesian MAP estimator can imply various complexity penalties, linear and $2k \ln(p/k)$-type penalties in particular, that gives an additional insight in motivation behind such types of penalties.

Consider the model (4.1), where the number of possible predictors p may be larger then the number of observations n. Recall that $r = rank(X)$ and we assume that any r columns of X are linearly independent.

Assume some prior on the model size $\pi(k) = P(|M| = k)$, where $\pi(k) > 0$, $k = 0, ..., r$ and $\pi(k) = 0$ for $k > r$ since the model becomes nonidentifiable when the number of its parameters is larger than the number of observations (see Section 4.2).

For any $k = 0, ..., r - 1$, assume all $\binom{p}{k}$ various models of size k to be equally likely, that is, conditionally on $|M| = k$,

$$P(M \mid |M| = k) = \binom{p}{k}^{-1}.$$

The case $k = r = rank(X)$ is slightly different. Although there are $\binom{p}{r}$ various sets of predictors of size r, all of them evidently result in the same estimator for the mean vector and, in this sense, are essentially undistinguishable and associated with a *single* (saturated) model. Hence, in this case, we set

$$P(M \mid |M| = r) = 1 \tag{4.7}$$

Finally, assume the normal prior on the unknown vector of k coefficients of the model M: $\beta_M \sim N_p(0, \gamma\sigma^2(D_M X'XD_M)^+)$, where $\gamma > 0$ and the diagonal indicator matrix D_M was defined in Section 4.2. This is a well-known conventional g-prior of Zellner (1986).

For the proposed hierarchical prior, a straightforward calculus yields the posterior probability of a model M of size $|M| = 0, ..., r - 1$:

$$P(M|y) \propto \pi(|M|) \binom{p}{|M|}^{-1} (1 + \gamma)^{-\frac{|M|}{2}} \exp\left\{\frac{\gamma}{\gamma + 1} \frac{y'XD_M(D_M X'XD_M)^+ D_M X'y}{2\sigma^2}\right\} \tag{4.8}$$

Finding the most likely model leads then to the following MAP model selection criterion:

$$y'XD_M(D_M X'XD_M)^+ D_M X'y + 2\sigma^2(1 + 1/\gamma) \ln\left\{\binom{p}{|M|}^{-1} \pi(|M|)(1 + \gamma)^{-\frac{|M|}{2}}\right\} \to \max_M$$

or, equivalently,

$$\|y - X\hat{\beta}_M\|^2 + 2\sigma^2(1 + 1/\gamma) \ln\left\{\binom{p}{|M|} \pi(|M|)^{-1}(1 + \gamma)^{\frac{|M|}{2}}\right\} \to \min_M, \tag{4.9}$$

which is of the general type (4.2) with the complexity penalty

$$Pen(k) = 2\sigma^2(1 + 1/\gamma)\ln\left\{\binom{p}{k}\pi(k)^{-1}(1+\gamma)^{\frac{k}{2}}\right\}, \quad k = 0,\ldots,r-1 \quad (4.10)$$

Similarly, for $|M| = r$ from (4.7) one has

$$Pen(r) = 2\sigma^2(1 + 1/\gamma)\ln\left\{\pi(r)^{-1}(1+\gamma)^{\frac{r}{2}}\right\} \quad (4.11)$$

In particular, the (truncated if $p > r$) binomial prior $B(p,\xi)$ corresponds to the prior assumption that the indicators d_{jM} are independent. The binomial prior yields the linear penalty $Pen(k) = 2\sigma^2(1 + 1/\gamma)k\ln(\sqrt{1+\gamma}(1-\xi)/\xi) \sim 2\sigma^2 k\ln(\sqrt{\gamma}(1-\xi)/\xi)$, $k = 1,\ldots,r-1$ for sufficiently large variance ratio γ. The AIC criterion corresponds then to $\xi \sim \sqrt{\gamma}/(e+\sqrt{\gamma})$, while $\xi \sim \sqrt{\gamma}/(p+\sqrt{\gamma})$ leads to the RIC criterion. These relations again confirm our previous arguments in Section 4.3 that RIC should be appropriate for sparse cases, where the size of the true (unknown) model is believed to be much less than the number of possible predictors, while AIC is suitable for dense cases, where they are of the same order. Any binomial prior or, equivalently, any linear penalty cannot "kill two birds with one stone".

On the other hand, there is a class of priors associated with the $2k\ln(p/k)$-type penalties. In particular, the (truncated) geometric distribution $\pi(k) \propto q^k$, $k = 1,\ldots,r$ yields $Pen(k) \sim 2\sigma^2(1 + 1/\gamma)k(\ln(p/k) + \zeta(\gamma,q))$, $k = 1,\ldots,r-1$, where we used that $k\ln(p/k) \leqslant \ln(\binom{p}{k}) < k(\ln(p/k) + 1)$ (see Lemma A1 of Abramovich et al., 2010). In addition, (4.11) implies $Pen(r) = 2\sigma^2 c(q,\gamma)r$ for some constant $c(q,\gamma) > 1$ that goes along the lines with the remark on the requirement on $Pen(r)$ for $2k\ln(p/k)$-type penalties in Section 4.4.

The Bayesian interpretation of the complexity penalized estimators can be also exploited for their computations. Generally, minimizing (4.2) requires an NP-hard combinatorial search over all possible models. To make computations for high-dimensional data feasible in practice, one typically applies either various greedy algorithms (e.g., forward selection) approximating the global solution in (4.2) by a stepwise sequence of local ones, or convex relaxation methods (e.g., Lasso (Tibshirani, 1996) and Dantzig selector (Candés & Tao, 2007) for linear penalties) replacing the original combinatorial problem by a related convex program. The proposed Bayesian approach allows one instead to use the Gibbs sampler to efficiently generate a sequence of models from the posterior distribution $P(M|\mathbf{y})$ in (4.8) (see, e.g. George & McCulloch, 1993 for more detail). The key point is that the relevant models with highest posterior probabilities will appear most frequently and can be easily identified even for a generated sample of a relatively small length.

4.6 Concluding Remarks

In this paper we considered model selection in Gaussian linear regression for high-dimensional data, where the number of possible predictors may be even larger than the number of available observations. The procedure is based on minimizing penalized least squares with a penalty on a model size. We discussed asymptotic properties of the resulting estimators corresponding to different types of penalties. Bayesian interpretation allows one to better understand the intuition behind various penalties and provides a natural tool for obtaining a wide class of estimators of this type.

We showed that any linear penalty, including widely used AIC, C_p, BIC and RIC, cannot be simultaneously minimax for both sparse and dense cases. Moreover, the same conclusions are valid for the well-known Lasso (Tibshirani, 1996) and Dantzig (Candés & Tao, 2007) estimators that for the optimally chosen tuning parameter, under nearly-orthogonality conditions similar to those considered in this paper, can achieve only the same sub-optimal rate $p_0 \ln p$ as RIC (Bickel, Ritov & Tsybakov, 2009). These results are, in fact, should not be surprising since both Lasso and Dantzig estimators are essentially based on convex relaxations of $|M| = \|\beta_M\|_0$ in the linear complexity penalty in (4.2) in order to replace the original combinatorial problem by a convex program (see also remarks in the conclusion of Section 4.5). Thus, Lasso approximates the l_0-norm $\|\beta_M\|_0$ by the the corresponding l_1-norm $\|\beta_M\|_1$. On the other hand, the nonlinear $2k \ln(p/k)$-type penalty adapts to both sparse and dense cases.

It is also interesting to note that, unlike model identification or coefficients estimation problems, where multicollinearity is a "curse", it may become a "blessing" for estimating the mean vector. One can exploit strong correlations between predictors to reduce the size of a model (hence, to decrease the variance) without paying much extra price in the bias.

Acknowledgements The work was supported by Israel Science Foundation (ISF), grant ISF-248/08. Valuable remarks of the two referees are gratefully acknowledged.

References

1. Abramovich, F., Benjamini, Y., Donoho, D., Johnstone, I.: Adapting to unknown sparsity by controlling the false discovery rate. Ann. Statist. **34**, 584–653 (2006)
2. Abramovich, F., Grinshtein, V.: MAP model selection in Gaussian regression. Electron. J. Statist. **4**, 932–949 (2010)
3. Abramovich, F., Grinshtein, V., Petsa, A., Sapatinas, T.: On Bayesian testimation and its application to wavelet thresholding. Biometrika **97**, 181–198 (2010)
4. Akaike, H.: Information theory and an extension of the maximum likelihood principle. In: B. Petrov, F. Czáki (eds.) Proceedings of the Second International Symposium on Information Theory, pp. 267–281. Akademiai Kiadó, Budapest (1973)
5. Bickel, P., Ritov, Y., Tsybakov, A.: Simultaneous analysis of Lasso and Dantzig selector. Ann. Statist. **35**, 1705–1732 (2009)
6. Birgé, L., Massart, P.: Gaussian model selection. J. Eur. Math. Soc. **3**, 203–268 (2001)
7. Birgé, L., Massart, P.: Minimal penalties for Gaussian model selection. Probab. Theory Related Fields **138**, 33–73 (2007)
8. Bunea, F., Tsybakov, A., Wegkamp, M.: Aggregation for Gaussian regression. Ann. Statist. **35**, 1674–1697 (2007)
9. Candés, E., Tao, T.: The Dantzig selector: statistical estimation when p is much larger than n. Ann. Statist. **35**, 2313–2351 (2007)

10. Foster, D., George, E.: The risk inflation criterion for multiple regression. Ann. Statist. **22**, 1947–1975 (1994)
11. George, E., McCuloch, R.: Variable selection via Gibbs sampling. J. Amer. Statist. Assoc. **88**, 881–889 (1993)
12. Mallows, C.: Some comments on C_p. Technometrics **15**, 661–675 (1973)
13. Schwarz, G.: Estimating the dimension of a model. Ann. Statist. **6**, 461–464 (1978)
14. Tibshirani, R.: Regression shrinkage and selection via the Lasso. J. R. Stat. Soc. Ser. B Stat. Methodol. **58**, 267–288 (1996)
15. Wu, Z., Zhou, H.: Model selection and sharp asymptotic minimaxity. Tech. rep., Statistics Department, Yale University (2010)
16. Zellner, A.: On assessing prior distributions and Bayesian regression analysis with g-prior distributions. In: P. Goel, A. Zellner (eds.) Bayesian Inference and Decision Techniques: Essays in Honor of Bruno de Finietti, pp. 233–243. North-Holland, Amsterdam (1986)

Chapter 5
Bayesian Perspectives on Sparse Empirical Bayes Analysis (SEBA)

Natalia Bochkina and Ya'acov Ritov

... and only a star or two set sparsedly in the vault of heaven; and you will find a sight as stimulating as the hoariest summit of the Alps. — R. L. Stevenson

Abstract We consider a joint processing of n independent similar sparse regression problems. Each is based on a sample $(y_{i1}, x_{i1}) \ldots, (y_{im}, x_{im})$ of m i.i.d. observations from $y_{i1} = x_{i1}^{\mathsf{T}} \beta_i + \varepsilon_{i1}$, $y_{i1} \in \mathbb{R}$, $x_{i1} \in \mathbb{R}^p$, and $\varepsilon_{i1} \sim N(0, \sigma^2)$, say. The dimension p is large enough so that the empirical risk minimizer is not feasible. We consider, from a Bayesian point of view, three possible extensions of the lasso. Each of the three estimators, the lassoes, the group lasso, and the RING lasso, utilizes different assumptions on the relation between the n vectors $\beta_1, \ldots, \beta_n$.

5.1 Introduction

We consider the model

$$y_{ij} = x_{ij}^{\mathsf{T}} \beta_i + \varepsilon_{ij}, \quad i = 1, \ldots, n, \ j = 1, \ldots, m,$$

or in a standard vector form

$$Y_i = X_i^{\mathsf{T}} \beta_i + \varepsilon_i, \quad i = 1, \ldots, n, \tag{5.1}$$

Natalia Bochkina
University of Edinburgh, School of Mathematics, King's Buildings, Mayfield Road, Edinburgh, EH9 3JZ, UK, e-mail: N.Bochkina@ed.ac.uk

Ya'acov Ritov
The Hebrew University of Jerusalem, Department of Statistics, Mount Scopus, Jerusalem 91905, Israel, e-mail: yaacov.ritov@gmail.com

where $\beta_i \in \mathbb{R}^p$. The matrix $X_i \in \mathbb{R}^{m \times p}$ is either deterministic fixed design matrix, or a sample of m independent $\mathbb{R}^p$ random vectors. Finally, ε_{ij}, $i = 1, \ldots, n$, $j = 1, \ldots, m$ are (at least uncorrelated with the x's), but typically assumed to be i.i.d. sub-Gaussian random variables, independent of the regressors x_{ij}. We can consider this as n partially related regression models, with m i.i.d. observations on the each model. For simplicity, we assume that all variables have expectation 0. The fact that the number of observations does not dependent on i is arbitrary and is assumed only for the sake of notational simplicity. Let $\mathscr{B}$ be the matrix $(\beta_1, \ldots, \beta_n)$.

The standard FDA (functional data analysis) is of this form, when the functions are approximated by their projections on some basis. Here we have n i.i.d. random functions, and each group can be considered as m noisy observations, each one is on the value of these functions at a given value of the argument. Thus,

$$y_{ij} = g_i(z_{ij}) + \varepsilon_{ij}, \tag{5.2}$$

where $z_{ij} \in [0,1]$. The model fits the regression setup of (5.1), if $g(z) = \sum_{\ell=1}^{p} \beta_\ell h_\ell(p)$ where $h_1, \ldots, h_p$ are in $L_2(0,1)$, and $x_{ij\ell} = h_\ell(z_{ij})$.

This approach is in the spirit of the empirical Bayes (compound decision) approach. Note however that the term "empirical Bayes" has a few other meanings in the literature, cf, [8, 9, 12]. The empirical Bayes approach to sparsity was considered before, e.g., [13, 3, 4, 6]. However, in these discussions the compound decision problem was within a single vector, while we consider the compound decision to be between the vectors, where the vectors are the basic units. The beauty of the concept of compound decision, is that we do not have to assume that in reality the units are related. They are considered as related only because our loss function is additive.

One of the standard tools for finding sparse solutions in a large p small m situation is the lasso (Tibshirani [10]), and the methods we consider are possible extensions.

We will make use of the following notation, introducing the $l_{p,q}$ norm of matrices and sets z of vectors:

Definition 5.1. For a matrix A, $\|A\|_{p,q} = \left(\sum_i \left(\sum_j A_{ij}^p \right)^{q/p} \right)^{1/q}$. If $z_1, \ldots, z_n$, is a collection of vectors, not necessarily of the same length, z_{ij}, $i = 1, \ldots, n$, $j = 1, \ldots, J_i$, then $\|\{z_1, \ldots, z_n\}\|_{p,q} = \left[\sum_{i=1}^{n} \left(\sum_{j \in J_i} |z_{ij}|^p \right)^{q/p} \right]^{1/q}$.

To simplify the notation, in this paper we will also use element-wise matrix norms $\|A\|_p = \|A\|_{p,p}$.

These norms will serve as a penalty on the size of the matrix $\mathscr{B} = (\beta_1, \ldots, \beta_n)$. Different norms imply different estimators, each appropriate under different assumptions.

Within the framework of the compound decision theory, we can have different scenarios. The first one is that the n groups are considered as repeated similar models for p variables, and the aim is to choose the variables that are useful for all models. The relevant variation of the lasso procedure in this case is the group lasso introduced by Yuan and Lin [11]:

$$\hat{\mathscr{B}} = \arg\min_{\mathscr{B}} \sum_{i=1}^{n} \sum_{j=1}^{m} (y_{ij} - x_{ij}^{\mathsf{T}}\beta_i)^2 + \lambda \|\mathscr{B}^{\mathsf{T}}\|_{2,1}. \tag{5.3}$$

Yuan and Lin also showed that in this case the sparsity pattern of variables (that is, the subset of variable with non-zero coefficients) is the same (with probability 1). Non-asymptotic inequalities under restricted eigenvalue type condition for group lasso are given by Lounici et al. [7].

Another possible scenario is where there is no direct relationship between the groups, and the only way the data are combined together is via the selection of the common penalty. In this case the sparsity pattern of the solution for each group are unrelated. We argue that the alternative formulation of the lasso procedure:

$$\hat{\mathscr{B}} = \arg\min_{\mathscr{B}} \sum_{i=1}^{n} \sum_{j=1}^{m} (y_{ij} - x_{ij}^{\mathsf{T}}\beta_i)^2 + \lambda \|\mathscr{B}\|_{1,\alpha}, \tag{5.4}$$

which we refer to as "lassoes" can be more natural than the simple lasso. The standard choice is $\alpha = 1$, but we believe that $\alpha > 4$ is, in fact, more consistent with a Bayesian point of view.

If we compare (5.4) to (5.3) we can see the difference between the prior assumptions of the lassoes and the group lasso. The basic elements of the lassoes are the β_i's vector, and we assume a priori that each of them is sparse. On the other hand, the basic elements of the group lasso are the variables, and we assume a priori that most of them do not contribute significantly to any of the regression equation.

We shall also consider a third situation where there is a sparse representation in some unknown basis, but assumed common to the n groups. The standard notion of sparsity, as captured by the ℓ_0 norm, or by the standard lasso, the lassoes, and the group lasso, is basis dependent. For example, when we prefer to leave it a priori open whether the function should be described in terms of the standard Haar wavelet basis, a collection of interval indicators, or a collection of step functions. All these three span the same linear space, but the true functions may be sparse in only one of them.

The rotation invariant group (RING) lasso was suggested as a natural extension of the group lasso to the situation where the proper sparse description of the regression function within a given basis is not known in advance ([2]). The corresponding penalty is the trace norm (or Schatten norm with $p = 1$) of the matrix $\mathscr{B}$, which finds the rotation that gives the best sparse representation of all vectors instantaneously.

The aim is to discuss the Bayesian interpretation of the three lasso extensions to the compound decision problem setting. Since the lassoes method, to our knowledge, has not been considered previously, we also present some theoretical results for it such as sparsity oracle inequalities and the persistency analysis (the latter is in the sense of [5], i.e., the equivalence of the empirical risk minimum and the correspondence population minimum. See Subsection 5.2.1).

The chapter is organised as follows. In Section 5.2 we introduce the lassoes method, discuss the Bayesian perspective, perform the persistency analysis and give the sparsity oracle inequalities. Section 5.3 is devoted to a Bayesian perspective on group lasso and Section 5.4 - to a Bayesian perspective on RING lasso. All the proofs are given in the Appendix.

5.2 The Lassoes Procedure

5.2.1 Persistency and Bayesian Interpretation

The minimal structural relationship we may assume is that the β's are not related, except that we believe that there is a bound on the average sparsity of the β's. One possible approach would be to consider the problem as a standard sparse regression problem with nm observations, a single vector of coefficients $\beta = (\beta_1^\mathsf{T}, \ldots, \beta_n^\mathsf{T})^\mathsf{T}$, and a block diagonal design matrix X. This solution, which corresponds to the solution of (5.4) with $\alpha = 1$, imposes very little on the similarity among $\beta_1, \ldots, \beta_n$. The lassoes procedure discussed in this section assumes that these vectors are similar, at least in their level of sparsity.

We assume that each vector of β_i, $i = 1, \ldots, n$, solves a different problem, and these problems are related only through the common penalty in the joint loss function, which is the sum of the individual losses, see (5.4).

We want to introduce some notation. We assume that for each $i = 1, \ldots, n$, $z_{ij} = (y_{ij}, x_{ij}^\mathsf{T})^\mathsf{T}$, $j = 1, \ldots, m$ are i.i.d., sub-Gaussian random variables, drawn from a distribution Q_i. Let $z_i = (y_i, x_i^\mathsf{T})^\mathsf{T}$ be an independent sample from Q_i. For any vector a, let $\tilde{a} = (-1, a^\mathsf{T})^\mathsf{T}$, and let $\tilde{\Sigma}_i$ be the covariance matrix of z_i and $\mathfrak{S} = (\tilde{\Sigma}_1, \ldots, \tilde{\Sigma}_n)$. The goal is to find the matrix $\hat{\mathscr{B}} = (\hat{\beta}_1, \ldots, \hat{\beta}_n)$ that minimizes the mean prediction error:

$$L(\mathscr{B}, \mathfrak{S}) = \sum_{i=1}^{n} \mathsf{e}_{Q_i}(y_i - x_i^\mathsf{T}\beta_i)^2 = \sum_{i=1}^{n} \tilde{\beta}_i^\mathsf{T} \tilde{\Sigma}_i \tilde{\beta}_i. \tag{5.5}$$

For p small, the natural approach is empirical risk minimization, that is replacing $\tilde{\Sigma}_i$ in (5.5) by $\tilde{S}_i$, the empirical covariance matrix of z_i. However, generally speaking, if p is large, empirical risk minimization results in overfitting the data. Greenshtein and Ritov [5] suggested (for the standard $n = 1$) minimization over a restricted set of possible β's, in particular, to either ℓ_1 or ℓ_0 balls. In fact, their argument is based on the following simple observations

$$\left| \tilde{\beta}^\mathsf{T} (\tilde{\Sigma}_i - \tilde{S}_i) \tilde{\beta} \right| \leqslant \|\tilde{\Sigma}_i - \tilde{S}_i\|_\infty \|\tilde{\beta}\|_1^2$$

and $\tag{5.6}$

$$\|\tilde{\Sigma}_i - \tilde{S}_i\|_\infty = \mathcal{O}_p(m^{-1/2}\log p),$$

where $\|A\|_\infty = \max_{i,j} |A_{ij}|$.

This leads to the natural extension of the single vector lasso to the compound decision problem set up, where we penalize by the sum of the *squared ℓ_1 norms* of vectors $\tilde{\beta}_1, \ldots, \tilde{\beta}_n$, and obtain the estimator defined by:

$$(\tilde{\beta}_i, \ldots, \tilde{\beta}_n) = \underset{\tilde{\beta}_1, \ldots, \tilde{\beta}_n}{\arg\min} \left\{ m \sum_{i=1}^{n} \tilde{\beta}_i^{\mathsf{T}} \tilde{S}_i \tilde{\beta}_i + \lambda_m \sum_{i=1}^{n} \|\tilde{\beta}_i\|_1^2 \right\}$$

$$= \underset{\tilde{\beta}_1, \ldots, \tilde{\beta}_n}{\arg\min} \sum_{i=1}^{n} \left\{ \sum_{j=1}^{m} (y_{ij} - x_{ij}^{\mathsf{T}} \beta_i)^2 + \lambda_m \|\tilde{\beta}_i\|_1^2 \right\}. \tag{5.7}$$

Note that when $n = 1$, the fact that we penalized by the squared ℓ_1 norm, and not by the ℓ_1 norm itself does not make a difference. To be more exact, if $n = 1$, for any λ_m in (5.7), there is λ_m' such that the least square with penalty $\lambda_m' \|\tilde{\beta}_1\|_1$ yields the same value as (5.7). For convenience we consider the asymptotic analysis as function of m. That is we consider $n = n_m$, $p = p_m$, and $\lambda = \lambda_m$. Often, the subscript m will be dropped.

Also, (5.7) may seem as, and numerically it certainly is, n separate problems, each involving a specific β_i. The problems are related however, because the penalty function is the same for all. They are tied, therefore, by the chosen value of λ_m, whether this is done *a-priori*, or by solving a single constraint maximization problem, or if λ_m is selected *a-posteriori* by a method like cross validation.

The prediction error of the lassoes estimator can be bounded in the following way. In the statement of the theorem, c_n is the minimal achievable risk, while C_n is the risk achieved by a particular sparse solution.

Theorem 5.1. *Let β_{i0}, $i = 1, \ldots, n$ be n arbitrary vectors and let $C_n = n^{-1} \sum_{i=1}^{n} \tilde{\beta}_{i0}^{\mathsf{T}} \tilde{\Sigma}_i \tilde{\beta}_{i0}$. Let $c_n = n^{-1} \sum_{i=1}^{n} \min_\beta \tilde{\beta}^{\mathsf{T}} \tilde{\Sigma}_i \tilde{\beta}$. Then*

$$\sum_{i=1}^{n} \tilde{\beta}_i^{\mathsf{T}} \tilde{\Sigma}_i \tilde{\beta}_i \leqslant \sum_{i=1}^{n} \tilde{\beta}_{i0}^{\mathsf{T}} \tilde{\Sigma}_i \tilde{\beta}_{i0} + \left(\frac{\lambda_m}{m} + \delta_m \right) \sum_{i=1}^{n} \|\tilde{\beta}_{i0}\|_1^2 - \left(\frac{\lambda_m}{m} - \delta_m \right) \sum_{i=1}^{n} \|\tilde{\beta}_i\|_1^2,$$

where $\delta_m = \max_i \|\tilde{S}_i - \tilde{\Sigma}_i\|_\infty$, and $(\tilde{\beta}_i, \ldots, \tilde{\beta}_n)$ are given in (5.7). If also $\lambda_m/m \to 0$ and $\lambda_m/(m^{1/2} \log(np)) \to \infty$ (and necessarily $m \to \infty$), then

$$\sum_{i=1}^{n} \|\tilde{\beta}_i\|_1^2 = \mathcal{O}_p\left(mn \frac{C_n - c_n}{\lambda_m} \right) + \left(1 + \mathcal{O}\left(\frac{m^{1/2}}{\lambda_m} \log(np) \right) \right) \sum_{i=1}^{n} \|\tilde{\beta}_{i0}\|_1^2 \tag{5.8}$$

and

$$\sum_{i=1}^{n} \tilde{\beta}_i^{\mathsf{T}} \tilde{\Sigma}_i \tilde{\beta}_i \leqslant \sum_{i=1}^{n} \tilde{\beta}_{i0}^{\mathsf{T}} \tilde{\Sigma}_i \tilde{\beta}_{i0} + (1 + o_p(1)) \frac{\lambda_m}{m} \sum_{i=1}^{n} \|\tilde{\beta}_{i0}\|_1^2.$$

The result is meaningful, although not as strong as may be wished, as long as $C_n - c_n \to 0$, while $n^{-1} \sum_{i=1}^{n} \|\tilde{\beta}_{i0}\|_1^2 = o_p(m^{1/2})$. That is, when there is a relatively sparse approximations to the best regression functions. Here sparse means only that the ℓ_1 norms of vectors is strictly smaller, on the average, than $\sqrt{m}$. Of course, if the minimizer of $\tilde{\beta}^{\mathsf{T}} \tilde{\Sigma}_i \tilde{\beta}$ itself is sparse, then by (5.8) $\tilde{\beta}_1, \ldots, \tilde{\beta}_n$ are as sparse as the true minimizers.

Also note, that the prescription that the theorem gives for selecting λ_m, is sharp: choose λ_m as close as possible to $m\delta_m$, or slightly larger than $\sqrt{m}$.

The estimators $\tilde{\beta}_1, \ldots, \tilde{\beta}_m$ look as if they are the mode of the a-posteriori distribution of the β_i's when $y_{ij}|\beta_i \sim N(x_{ij}^\mathsf{T}\beta_i, \sigma^2)$, the $\beta_1, \ldots, \beta_n$ are a priori independent, and β_i has a prior density proportional to $\exp(-\lambda_m\|\tilde{\beta}_i\|_1^2/\sigma^2)$. This distribution can be constructed as follows. Suppose $T_i \sim N(0, \lambda_m^{-1}\sigma^2)$. Given T_i, let $u_{i1}, \ldots, u_{ip}$ be distributed uniformly on the simplex $\{u_{i\ell} \geqslant 0, \sum_{\ell=1}^n u_{i\ell} = |T_i|\}$. Let $s_{i1}, \ldots, s_{ip}$ be i.i.d. Rademacher random variables (taking values ± 1 with probabilities 0.5), independent of $T_i, u_{i1}, \ldots, u_{ip}$. Finally let $\beta_{i\ell} = u_{i\ell}s_{i\ell}, \ell = 1, \ldots, p$.

However, this Bayesian point of view is not consistent with the above suggested value of λ_m. An appropriate prior should express the beliefs on the unknown parameter which are by definition conceptually independent of the amount data to be collected. However, the permitted range of λ_m does not depend on the assumed range of $\|\tilde{\beta}_i\|$, but quite artificially should be in order between $m^{1/2}$ and m. That is, the penalty should be increased with the number of observations on each β_i, although at a slower rate than m. In fact, even if we relax what we mean by "prior", the value of λ_m goes in the 'wrong' direction. As $m \to \infty$, one may wish to use weaker a-priori assumptions, and allow T above to have a-priori second moment going to infinity, not to 0, as entailed by $\lambda_m \to 0$.

The Bayesian inconsistency does not come from the asymptotic setup, and it does not come from considering a more and more complex model. It was presented as asymptotic in $m \to \infty$, because it is clear that asymptotically we get the wrong results, but the phenomena occurs along the way and not only in the final asymptotic destination. The parameter λ should be much larger than we believe *a priori* that $\|\tilde{\beta}_i\|_1^{-1}$ should be. If λ is chosen such that the prior distribution have the level of sparsity we believe in, then the posteriori distribution would not be sparse at all! To obtain a sparse solution, we should pose a prior which predicts an almost zero vector β. Also, the problem does not follow from increasing the dimension, because the asymptotic is in m and not in p, the latter is very large along the process. We could start the "asymptotic" discussion with m_0 observations per $\beta_i \in \mathbb{R}^{p_0}$, p_0 almost exponential in m_0. Then we could keep p constant, while increasing m. We would get the inconsistency much before m will be $\mathcal{O}(p_0)$. Finally, the Bayesian inconsistency is not because the real dimension, the number of non-zero entries of β_i, is increasing. In fact, the inconsistency appears when this number is kept of the same order, and the prior predicts increasingly sparse vectors (but not fast enough). In short, the problem is that the considered prior distribution cannot compete with the likelihood when the dimension of the observations is large (note, just 'large', not 'asymptotically large').

We would like to consider a more general penalty of the form $\sum_{i=1}^n \|\beta_i\|_1^\alpha$. A power $\alpha \neq 1$ of ℓ_1 norm of β as a penalty introduces a priori dependence between the variables which is not the case for the regular lasso penalty with $\alpha = 1$, where all β_{ij} are a priori independent. As α increases, the sparsity of the different vectors tends to be the same—the price for a single non-sparse vector is higher as α increases. Note that given the value of λ_m, the n problems are treated independently. The compound decision problem is reduced to picking a common level of penalty. When this choice is data based, the different vectors become dependent. This is the main benefit of this approach—the selection of the regularization is based on all the mn observations.

For a proper Bayesian perspective, we need to consider a prior with much smaller tails than the normal. Suppose for simplicity that $c_n = C_n$ (that is, the "true" regressors are sparse), and $\max_i \|\beta_{i0}\|_1 < \infty$.

Theorem 5.2. *Let β_{i0} be the minimizer of $\tilde{\beta}^\top \Sigma_i \tilde{\beta}$. Suppose $\max_i \|\beta_{i0}\|_1 < \infty$. Consider the estimators:*

$$(\tilde{\beta}_i, \ldots, \tilde{\beta}_n) = \underset{\tilde{\beta}_1, \ldots, \tilde{\beta}_n}{\arg\min} \left\{ m \sum_{i=1}^n \tilde{\beta}_i^\top \tilde{S}_i \tilde{\beta}_i + \lambda_m \sum_{i=1}^n \|\tilde{\beta}_i\|_1^\alpha \right\}$$

for some $\alpha > 2$. Assume $m \to \infty$ and $\lambda_n = \mathcal{O}(m\delta_m) = \mathcal{O}(m^{1/2}\log p)$. Then

$$n^{-1} \sum_{i=1}^n \|\tilde{\beta}_i\|_1^2 = \mathcal{O}((m\delta_m/\lambda_m)^{2/(\alpha-2)}),$$

and

$$\sum_{i=1}^n \tilde{\beta}_i^\top \Sigma_i \tilde{\beta}_i \leqslant \sum_{i=1}^n \tilde{\beta}_{i0}^\top \Sigma_i \tilde{\beta}_{i0} + \mathcal{O}_p(n(m/\lambda_m)^{2/(\alpha-2)}\delta_m^{\alpha/(\alpha-2)}).$$

Note that, if we, the Bayesians, believe that $n^{-1}\sum_{i=1}^n \|\tilde{\beta}_i\|_1^2 \overset{p}{\longrightarrow} c$, then λ_m should also converge to a constant. Theorem 5.2 implies that the estimator is persistent (i.e., the empirical minimization gives the same risk as the population minimizer) if $m^2\delta_n^\alpha \to 0$, or $\alpha > 4$. That is, the prior should have a very short tails. In fact, if the prior's tails are short enough, we can accommodate an increasing value of the $\tilde{\beta}_i$'s by taking $\lambda_m \to 0$.

The theorem suggests a simple way to select λ_m based on the data. Note that $n^{-1}\sum_{i=1}^n \|\tilde{\beta}_i\|_1^2$ is a decreasing function of λ. Hence, we can start with a very large value of λ and decrease it until $n^{-1}\sum_{i=1}^n \|\tilde{\beta}_i\|_1^2 \approx \lambda^{-2/\alpha}$.

We want to conclude on another role of the parameter α. The parameter λ_m controls the average $n^{-1}\sum_{i=1}^n \|\tilde{\beta}_i\|_1^\alpha$. When $\alpha = 1$, we are relatively flexible and allow some $\|\tilde{\beta}_i\|_1$ to be very large, as long as other are small. If α is larger, the penalty for $\|\tilde{\beta}_i\|_1$ much larger than the average becomes too large, and the solution tends to be with all $\|\tilde{\beta}_i\|_1$ being of the same order.

5.2.2 Restricted Eigenvalues Condition and Oracle Inequalities

The above discussion was based on the persistent type of argument. The results are relatively weak, but in return the conditions are very weak. For completeness we give much stronger results based on much stronger conditions. We show that the needed coefficient of the penalty function remains the same, and therefore the

Bayesian discussion did not depend on the results presented above. Before stating the conditions and the resulted inequalities we introduce some notation and definitions.

For a vector β, let $\mathcal{M}(\beta)$ be the cardinality of its support: $\mathcal{M}(\beta) = \sum_i \mathbf{I}(\beta_i \neq 0)$. Given a matrix $\Delta \in \mathbb{R}^{n \times p}$ and given a set $J = \{J_i\}$, $J_i \subset \{1, \ldots, p\}$, we denote $\Delta_J = \{\Delta_{i,j}, i = 1, \ldots, n, j \in J_i\}$. By the complement J^c of J we denote the set $\{J_1^c, \ldots, J_n^c\}$, i.e. the set of complements of J_i's. Below, X is $np \times m$ block diagonal design matrix, $X = \mathrm{diag}(X_1, X_2, \ldots, X_n)$, and with some abuse of notation, a matrix $\Delta = (\Delta_1, \ldots, \Delta_n)$ may be considered as the vector $(\Delta_1^\mathsf{T}, \ldots, \Delta_n^\mathsf{T})^\mathsf{T}$, and accordingly the norms $\|\Delta\|_\alpha$ are defined. Finally, recall the notation $\mathscr{B} = (\beta_1, \ldots, \beta_n)$

The restricted eigenvalue assumption of Bickel et al. [1] (and Lounici et al. [7]) can be generalized to incorporate unequal subsets J_is. In the assumption below, the restriction is given in terms of $\ell_{q,1}$ norm, $q \geqslant 1$.

Assumption $\mathrm{RE}_q(s, c_0, \kappa)$.

$$\kappa = \min\left\{ \frac{\|X^\mathsf{T}\Delta\|_2}{\sqrt{m}\|\Delta_J\|_2} : \max_i |J_i| \leqslant s, \Delta \in \mathbb{R}^{n \times p} \setminus \{0\}, \|\Delta_{J^c}\|_{q,1} \leqslant c_0 \|\Delta_J\|_{q,1} \right\} > 0.$$

We apply it with $q = 1$, and in Lounici et al. [7] it was used for $q = 2$. We call it a *restricted eigenvalue assumption* to be consistent with the literature. In fact, as stated it is a definition of κ as the maximal value that satisfies the condition, and the only real assumption is that κ is positive. However, the larger κ is, the more useful the "assumption" is. Discussion of the normalisation by $\sqrt{m}$ can be found in Lounici et al. [7].

For penalty $\lambda \sum_i \|\beta_i\|_1^\alpha$, we have the following inequalities.

Theorem 5.3. *Assume $y_{ij} \sim \mathcal{N}(x_{ij}^\mathsf{T}\beta_i, \sigma^2)$, and let $\hat{\mathscr{B}} = (\hat{\beta}_1, \ldots, \hat{\beta}_n)$ correspond to a minimizer of (5.7), with*

$$\lambda = \lambda_m \geqslant \frac{4A\sigma\sqrt{m\log(np)}}{\alpha \max(B^{\alpha-1}, \hat{B}^{\alpha-1})},$$

where $\alpha \geqslant 1$ and $A > \sqrt{2}$, $B \geqslant \max_i \|\beta_i\|_1$ and $\hat{B} \geqslant \max_i \|\hat{\beta}_i\|_1$, $\max(B, \hat{B}) > 0$ (B may depend on n, m, p, and so can $\hat{B}$). Suppose that generalized assumption $RE_1(s, 3, \kappa)$ defined above holds, $\sum_{j=1}^m x_{ij\ell}^2 = m$ for all i, ℓ, and $\mathcal{M}(\beta_i) \leqslant s$ for all i.

Then, with probability at least $1 - (np)^{1-A^2/2}$,

(a) The root means squared prediction error is bounded by:

$$\frac{1}{\sqrt{nm}}\|X^\mathsf{T}(\hat{\beta} - \beta)\|_2 \leqslant \frac{\sqrt{s}}{\kappa\sqrt{m}}\left[\frac{3\alpha\lambda}{2\sqrt{m}}\max(B^{\alpha-1}, \hat{B}^{\alpha-1}) + 2A\sigma\sqrt{\log(np)}\right],$$

(b) The mean estimation absolute error is bounded by:

$$\frac{1}{n}\|\mathscr{B} - \hat{\mathscr{B}}\|_1 \leqslant \frac{4s}{m\kappa^2}\left[\frac{3\alpha\lambda}{2}\max(B^{\alpha-1}, \hat{B}^{\alpha-1}) + 2A\sigma\sqrt{m\log(np)}\right],$$

(c)

$$\mathscr{M}(\hat{\beta}_i) \leqslant \|X_i^{\mathsf{T}}(\beta_i - \hat{\beta}_i)\|_2^2 \frac{m\phi_{i,\max}}{\left(\lambda\alpha\|\hat{\beta}_i\|_1^{\alpha-1}/2 - A\sigma\sqrt{m\log(np)}\right)^2},$$

where $\phi_{i,\max}$ is the maximal eigenvalue of $X_i^{\mathsf{T}}X_i/m$, and β and $\hat{\beta}$ are vector forms of matrices $\mathscr{B}$ and $\hat{\mathscr{B}}$ respectively.

Note that for $\alpha = 1$, if we take $\lambda = 2A\sigma\sqrt{m\log(np)}$, the bounds are of the same order as for the lasso with np-dimensional β (up to a constant of 2, cf. Theorem 7.2 in Bickel et al. [1]). For $\alpha > 1$, we have dependence of the bounds on the ℓ_1 norm of β and $\hat{\beta}$.

We can use bounds on the norm of $\hat{\beta}$ given in Theorem 5.2 to obtain the following results.

Theorem 5.4. *Assume* $y_{ij} \sim \mathscr{N}(x_{ij}^{\mathsf{T}}\beta_i, \sigma^2)$, *with* $\max_i\|\beta_i\|_1 \leqslant b$ *where* $b > 0$ *can depend on* n, m, p. *Take some* $\eta \in (0,1)$. *Let* $\hat{\mathscr{B}} = (\hat{\beta}_1, \ldots, \hat{\beta}_n)$ *correspond to a minimizer of* (5.7), *with*

$$\lambda = \lambda_m = \frac{4A\sigma}{\alpha b^{\alpha-1}}\sqrt{m\log(np)},$$

$A > \sqrt{2}$, *such that* $b > c\eta^{1/(2(\alpha-1))}$ *for some constant* $c > 0$. *Also, assume that* $C_n - c_n = \mathscr{O}(m\delta_n)$, *as defined in Theorem 5.1.*

Suppose that generalized assumption $RE_1(s, 3, \kappa)$ *defined above holds,* $\sum_{j=1}^m x_{ij\ell}^2 = m$ *for all* i, ℓ, *and* $\mathscr{M}(\beta_i) \leqslant s$ *for all* i.

Then, for some constant $C > 0$, *with probability at least* $1 - \left(\eta + (np)^{1-A^2/2}\right)$,

(a) The prediction error can be bounded by:

$$\|X^{\mathsf{T}}(\hat{\beta} - \beta)\|_2^2 \leqslant \frac{4A^2\sigma^2 sn\log(np)}{\kappa^2}\left[1 + 3C\left(\frac{b}{\sqrt{\eta}}\right)^{(\alpha-1)/(\alpha-2)}\right]^2,$$

(b) The estimation absolute error is bounded by:

$$\|\mathscr{B} - \hat{\mathscr{B}}\|_1 \leqslant \frac{2A\sigma sn\sqrt{\log(np)}}{\kappa^2\sqrt{m}}\left[1 + 3C\left(\frac{b}{\sqrt{\eta}}\right)^{(\alpha-1)/(\alpha-2)}\right].$$

(c) Average sparsity of $\hat{\beta}_i$:

$$\frac{1}{n}\sum_{i=1}^n \mathscr{M}(\hat{\beta}_i) \leqslant s\frac{4\phi_{\max}}{\kappa^2\delta^2}\left[1 + 3C\left(\frac{b}{\sqrt{\eta}}\right)^{1+1/(\alpha-2)}\right]^2,$$

where $\phi_{\max}$ is the largest eigenvalue of $X^{\mathsf{T}}X/m$.

This theorem also tells us how large ℓ_1 norm of β can be to ensure good bounds on the prediction and estimation errors.

Note that under the Gaussian model and fixed design matrix, assumption $C_n - c_n = \mathcal{O}(m\delta_n)$ is equivalent to $\|\mathscr{B}\|_2^2 \leqslant Cm\delta_n$.

5.3 Group Lasso: Bayesian Perspective

Write $\mathscr{B} = (\beta_1, \ldots, \beta_n) = (\mathfrak{b}_1^{\mathsf{T}}, \ldots, \mathfrak{b}_p^{\mathsf{T}})^{\mathsf{T}}$. The group lasso is defined (see Yuan and Lin [11]) by

$$\hat{\mathscr{B}} = \arg\min \left[\sum_{i=1}^{n} \sum_{j=1}^{m} (y_{ij} - x_{ij}^{\mathsf{T}} \beta_i)^2 + \lambda \sum_{\ell=1}^{p} \|\mathfrak{b}_\ell\|_2 \right] \qquad (5.9)$$

Note that $(\hat{\beta}_1, \ldots, \hat{\beta}_n)$ are defined as the minimum point of a strictly convex function, and hence they can be found by equating the gradient of this function to 0.

Note that (5.9) is equivalent to the mode of the a-posteriori distribution when given $\mathscr{B}$, Y_{ij}, $i = 1, \ldots, n$, $j = 1, \ldots, m$, are all independent, $y_{ij} \mid \mathscr{B} \sim \mathcal{N}(x_{ij}^{\mathsf{T}} \beta_i, \sigma^2)$, and a-priori, $\mathfrak{b}_1, \ldots, \mathfrak{b}_p$, are i.i.d.,

$$f_{\mathfrak{b}}(\mathfrak{b}_\ell) \propto \exp\{-\tilde{\lambda}\|\mathfrak{b}_\ell\|_2\}, \quad \ell = 1, \ldots, p,$$

where $\tilde{\lambda} = \lambda/(2\sigma^2)$. We consider now some property of this prior. For each ℓ, $\mathfrak{b}_\ell$ have a spherically symmetric distribution. In particular its components are uncorrelated and have mean 0. However, they are not independent. Change of variables to a polar system where

$$R_\ell = \|\mathfrak{b}_\ell\|_2$$
$$\beta_{\ell i} = R_\ell w_{\ell i}, \qquad w_\ell \in \mathbb{S}^{n-1},$$

where $\mathbb{S}^{n-1}$ is the sphere in $\mathbb{R}^n$. Then, clearly,

$$f(R_\ell, w_\ell) = C_{n,\lambda} R_\ell^{n-1} e^{-\tilde{\lambda} R_\ell}, \qquad R_\ell > 0, \qquad (5.10)$$

where $C_{n,\lambda} = \tilde{\lambda}^n \Gamma(n/2)/2\Gamma(n)\pi^{n/2}$. Thus, R_ℓ, w_ℓ are independent $R_\ell \sim \Gamma(n, \tilde{\lambda})$, and w_ℓ is uniform over the unit sphere.

The conditional distribution of one of the coordinates of $\mathfrak{b}_\ell$, say the first, given the rest has the form

$$f(\mathfrak{b}_{\ell 1} \mid \mathfrak{b}_{\ell 2}, \ldots, \mathfrak{b}_{\ell n}, \sum_{i=2}^{n} \mathfrak{b}_{\ell i}^2 = \rho^2) \propto e^{-\tilde{\lambda}\rho\sqrt{1+\mathfrak{b}_{\ell 1}^2/\rho^2}}$$

which for small $\mathfrak{b}_{\ell 1}/\rho$ looks like the normal density with mean 0 and variance $\rho/\tilde{\lambda}$, while for large $\mathfrak{b}_{\ell 1}/\rho$ behaves like the exponential distribution with mean $\tilde{\lambda}^{-1}$.

The sparsity property of the prior comes from the linear component of log-density of R. If $\tilde{\lambda}$ is large and the Ys are small, this component dominates the log-a-posteriori distribution and hence the maximum will be at 0.

Fix now $\ell \in \{1,\dots,p\}$, and consider the estimating equation for b_ℓ — the ℓ components of the β's. Fix the rest of the parameters and let $\tilde{Y}_{ij\ell}^{\mathscr{B}} = y_{ij} - \sum_{k \ne \ell} \beta_{ik} x_{ijk}$. Then $\hat{\mathsf{b}}_{\ell i}$, $i = 1,\dots,n$, satisfy

$$0 = -\sum_{j=1}^{m} x_{ij\ell}(\tilde{Y}_{ij\ell}^{\mathscr{B}} - \hat{\mathsf{b}}_{\ell i} x_{ij\ell}) + \frac{\lambda \hat{\mathsf{b}}_{\ell i}}{\sqrt{\sum_k \hat{\mathsf{b}}_{\ell k}^2}}, \qquad i = 1,\dots,n$$

$$= -\sum_{j=1}^{m} x_{ij\ell}(\tilde{Y}_{ij\ell}^{\mathscr{B}} - \hat{\mathsf{b}}_{\ell i} x_{ij\ell}) + \lambda_\ell^* \hat{\mathsf{b}}_{\ell i}, \qquad \text{say.}$$

Hence

$$\hat{\mathsf{b}}_{\ell i} = \frac{\sum_{j=1}^{m} x_{ij\ell} \tilde{Y}_{ij\ell}^{\mathscr{B}}}{\lambda_\ell^* + \sum_{j=1}^{m} x_{ij\ell}^2}. \tag{5.11}$$

The estimator has an intuitive appeal. It is the ordinary least square estimator of $\mathsf{b}_{\ell i}$, $\sum_{j=1}^{m} x_{ij\ell} \tilde{Y}_{ij\ell}^{\mathscr{B}} / \sum_{j=1}^{m} x_{ij\ell}^2$, pulled to 0. It is pulled less to zero as the variance of $\mathsf{b}_{\ell 1},\dots,\mathsf{b}_{\ell n}$ increases (and λ_ℓ^* is getting smaller), and as the variance of the LS estimator is lower (i.e., when $\sum_{j=1}^{m} x_{ij\ell}^2$ is larger).

If the design is well balanced, $\sum_{j=1}^{m} x_{ij\ell}^2 \equiv m$, then we can characterize the solution as follows. For a fixed ℓ, $\hat{\mathsf{b}}_{\ell 1},\dots,\hat{\mathsf{b}}_{\ell n}$ are the least square solution shrunk toward 0 by the same amount, which depends only on the estimated variance of $\hat{\mathsf{b}}_{\ell 1},\dots,\hat{\mathsf{b}}_{\ell n}$. In the extreme case, $\hat{\mathsf{b}}_{\ell 1} = \dots = \hat{\mathsf{b}}_{\ell n} = 0$, otherwise (assuming the error distribution is continuous) they are shrunken toward 0, but are different from 0.

We can use (5.11) to solve for λ_ℓ^*

$$\left(\frac{\lambda}{\lambda_\ell^*}\right)^2 = \|\hat{\mathsf{b}}_\ell\|_2^2 = \sum_{i=1}^{n} \left(\frac{\sum_{j=1}^{m} x_{ij\ell} \tilde{Y}_{ij\ell}^{\mathscr{B}}}{\lambda_\ell^* + \sum_{j=1}^{m} x_{ij\ell}^2}\right)^2.$$

Hence λ_ℓ^* is the solution of

$$\lambda^2 = \sum_{i=1}^{n} \left(\frac{\lambda_\ell^* \sum_{j=1}^{m} x_{ij\ell} \tilde{Y}_{ij\ell}^{\mathscr{B}}}{\lambda_\ell^* + \sum_{j=1}^{m} x_{ij\ell}^2}\right)^2. \tag{5.12}$$

Note that the RHS is monotone increasing, so (5.12) has at most a unique solution. It has no solution if at the limit $\lambda_\ell^* \to \infty$, the RHS is still less than λ^2. That is if

$$\lambda^2 > \sum_{i=1}^{n} \left(\sum_{j=1}^{m} x_{ij\ell} \tilde{Y}_{ij\ell}^{\mathscr{B}}\right)^2$$

then $\hat{\mathsf{b}}_\ell = 0$. In particular if

$$\lambda^2 > \sum_{i=1}^{n} \left(\sum_{j=1}^{m} x_{ij\ell} Y_{ij\ell} \right)^2, \qquad \ell = 1, \dots, p$$

Then all the random effect vectors are 0. In the balanced case the RHS is $\mathcal{O}_p(mn \log(p))$. By (5.10), this means that if we want that the estimator will be 0 if the underlined true parameters are 0, then the prior should prescribe that $\mathfrak{b}_\ell$ has norm which is $\mathrm{o}(m^{-1})$. This conclusion is supported by the recommended value of λ given, e.g. in [7].

5.4 RING Lasso: Bayesian Perspective

Let $A = \sum c_i x_i x_i^\mathsf{T}$, be a positive semi-definite matrix, where $x_1, x_2, \dots$ is an orthonormal basis of eigenvectors. Then, we define $A^\gamma = \sum c_i^\gamma x_i x_i^\mathsf{T}$. We consider now as penalty the function

$$|||\mathscr{B}|||_1 = \mathrm{trace}\left\{ \left(\sum_{i=1}^{n} \beta_i \beta_i^\mathsf{T} \right)^{1/2} \right\},$$

where $\mathscr{B} = (\beta_1, \dots, \beta_n) = (\mathfrak{b}_1^\mathsf{T}, \dots, \mathfrak{b}_p^\mathsf{T})^\mathsf{T}$. This is also known as trace norm or Schatten norm with $p = 1$. Note that $|||\mathscr{B}|||_1 = \sum c_i^{1/2}$ where $c_1, \dots, c_p$ are the eigenvalues of $\mathscr{B}\mathscr{B}^\mathsf{T} = \sum_{i=1}^{n} \beta_i \beta_i^\mathsf{T}$ (including multiplicities), i.e. this is the ℓ_1 norm on the singular values of $\mathscr{B}$. $|||\mathscr{B}|||_1$ is a convex function of $\mathscr{B}$.

In this section we study the estimator defined by

$$\hat{\mathscr{B}} = \underset{\mathscr{B} \in \mathbb{R}^{p \times n}}{\arg\min} \left\{ \sum_{i=1}^{n} \sum_{j=1}^{m} (y_{ij} - x_{ij}^\mathsf{T} \beta_i)^2 + \lambda |||\mathscr{B}|||_1. \right\} \tag{5.13}$$

We refer to this problem as RING (Rotation INvariant Group) lasso. See [2] for more details.

We consider now the penalty for β_k for a fixed k. Let $A = n^{-1} \sum_{k \neq i} \beta_k \beta_k^\mathsf{T}$, and write the spectral value decomposition $n^{-1} \sum_{k=1}^{n} \beta_k \beta_k^\mathsf{T} = \sum c_j x_j x_j^\mathsf{T}$ where $\{x_j\}$ is an orthonormal basis of eigenvectors. Using Taylor expansion for not too big β_i, we get

$$\mathrm{trace}\left((nA + \beta_i \beta_i^\mathsf{T})^{1/2} \right) \approx \sqrt{n}\, \mathrm{trace}(A^{1/2}) + \sum_{j=1}^{p} \frac{x_j^\mathsf{T} \beta_i \beta_i^\mathsf{T} x_j}{2 c_j^{1/2}}$$

$$= \sqrt{n}\, \mathrm{trace}(A^{1/2}) + \frac{1}{2} \beta_i^\mathsf{T} \left(\sum c_j^{-1/2} x_j x_j^\mathsf{T} \right) \beta_i$$

$$= \sqrt{n}\, \mathrm{trace}(A^{1/2}) + \frac{1}{2} \beta_i^\mathsf{T} A^{-1/2} \beta_i$$

Hence the estimator is as if β_i has a prior of $\mathcal{N}(0, n\sigma^2/\lambda A^{1/2})$. Note that the prior is only related to the estimated variance of β, and A appears with the power of $1/2$. Now A is not really the estimated variance of β, only the variance of the estimates, hence it should be inflated, and the square root takes care of that. Finally, note that eventually, if β_i is very large relative to nA, then the penalty becomes $\|\beta\|$, so the "prior" looks like normal for the center of the distribution and has exponential tails.

A better way to look on the penalty from a Bayesian perspective is to consider it as prior on the $n \times p$ matrix $\mathcal{B} = (\beta_1, \ldots, \beta_n)$. Recall that the penalty is invariant to the rotation of the matrix $\mathcal{B}$. In fact, $\|\|\mathcal{B}\|\|_1 = \|\|T\mathcal{B}U\|\|_1$, where T and U are $n \times n$ and $p \times p$ rotation matrices. Now, this means that if $\mathfrak{b}_1, \ldots, \mathfrak{b}_p$ are orthonormal set of eigenvectors of $\mathcal{B}^T\mathcal{B}$ and $\gamma_{ij} = \mathfrak{b}_j^T\beta_i$ — the PCA of $\beta_1, \ldots, \beta_n$, then $\|\|\mathcal{B}\|\|_1 = \sum_{j=1}^{p}\left(\sum_{i=1}^{n}\gamma_{ij}^2\right)^{1/2}$ — the RING lasso penalty in terms of the principal components. The "prior" is then proportional to $e^{-\tilde{\lambda}\sum_{j=1}^{p}\|\gamma_{\cdot j}\|_2}$ where $\tilde{\lambda} = \lambda/(2\sigma^2)$. Namely, we can obtain a random $\mathcal{B}$ from the prior by the following procedure:

1. Sample $r_1, \ldots, r_p$ independently from $\Gamma(n, \tilde{\lambda})$ distribution.
2. For each $j = 1, \ldots, p$ sample $\gamma_{1j}, \ldots, \gamma_{nj}$ independently and uniformly on the sphere with radius r_j.
3. Sample an orthonormal base $\chi_1, \ldots, \chi_p$ "uniformly".
4. Construct $\beta_i = \sum_{j=1}^{p}\gamma_{ij}\chi_j$.

Appendix

Proof (Theorem 5.1). Note that by the definition of $\tilde{\beta}_i$ and (5.6).

$$
\begin{aligned}
mnc_n + \lambda_m \sum_{i=1}^{n} \|\tilde{\beta}_i\|_1^2 \\
\leqslant m \sum_{i=1}^{n} \tilde{\beta}_i^T \tilde{\Sigma}_i \tilde{\beta}_i + \lambda_m \sum_{i=1}^{n} \|\tilde{\beta}_i\|_1^2 \\
\leqslant m \sum_{i=1}^{n} \tilde{\beta}_i^T \tilde{S}_i \tilde{\beta}_i + (\lambda_m + m\delta_m) \sum_{i=1}^{n} \|\tilde{\beta}_i\|_1^2 \\
\leqslant m \sum_{i=1}^{n} \tilde{\beta}_{i0}^T \tilde{S}_i \tilde{\beta}_{i0} + \lambda_m \sum_{i=1}^{n} \|\tilde{\beta}_{i0}\|_1^2 + m\delta_m \sum_{i=1}^{n} \|\tilde{\beta}_i\|_1^2 \\
\leqslant m \sum_{i=1}^{n} \tilde{\beta}_{i0}^T \tilde{\Sigma}_i \tilde{\beta}_{i0} + (\lambda_m + m\delta_m) \sum_{i=1}^{n} \|\tilde{\beta}_{i0}\|_1^2 + m\delta_m \sum_{i=1}^{n} \|\tilde{\beta}_i\|_1^2 \\
= mnC_n + (\lambda_m + m\delta_m) \sum_{i=1}^{n} \|\tilde{\beta}_{i0}\|_1^2 + m\delta_m \sum_{i=1}^{n} \|\tilde{\beta}_i\|_1^2.
\end{aligned}
\tag{5.14}
$$

Comparing the LHS with the RHS of (5.14), noting that $m\delta_m \ll \lambda_m$:

$$\sum_{i=1}^{n} \|\tilde{\tilde{\beta}}_i\|_1^2 \leqslant mn \frac{C_n - c_n}{\lambda_m - m\delta_m} + \frac{\lambda_m + m\delta_m}{\lambda_m - m\delta_m} \sum_{i=1}^{n} \|\tilde{\beta}_{i0}\|_1^2.$$

By (5.6) and (5.7):

$$
\begin{aligned}
\sum_{i=1}^{n} \tilde{\tilde{\beta}}_i^{\mathsf{T}} \tilde{\Sigma}_i \tilde{\tilde{\beta}}_i &\leqslant \sum_{i=1}^{n} \tilde{\tilde{\beta}}_i^{\mathsf{T}} \tilde{S}_i \tilde{\tilde{\beta}}_i + \delta_m \sum_{i=1}^{n} \|\tilde{\tilde{\beta}}_i\|_1^2 \\
&\leqslant \sum_{i=1}^{n} \tilde{\beta}_{i0}^{\mathsf{T}} \tilde{S}_i \tilde{\beta}_{i0} + \frac{\lambda_m}{m} \sum_{i=1}^{n} \|\tilde{\beta}_{i0}\|_1^2 - \frac{\lambda_m}{m} \sum_{i=1}^{n} \|\tilde{\tilde{\beta}}_i\|_1^2 + \delta_m \sum_{i=1}^{n} \|\tilde{\tilde{\beta}}_i\|_1^2 \\
&\leqslant \sum_{i=1}^{n} \tilde{\beta}_{i0}^{\mathsf{T}} \tilde{\Sigma}_i \tilde{\beta}_{i0} + \left(\frac{\lambda_m}{m} + \delta_m\right) \sum_{i=1}^{n} \|\tilde{\beta}_{i0}\|_1^2 - \left(\frac{\lambda_m}{m} - \delta_m\right) \sum_{i=1}^{n} \|\tilde{\tilde{\beta}}_i\|_1^2 \\
&\leqslant \sum_{i=1}^{n} \tilde{\beta}_{i0}^{\mathsf{T}} \tilde{\Sigma}_i \tilde{\beta}_{i0} + \left(\frac{\lambda_m}{m} + \delta_m\right) \sum_{i=1}^{n} \|\tilde{\beta}_{i0}\|_1^2.
\end{aligned}
\tag{5.15}
$$

The result follows.

Proof (Theorem 5.2). The proof is similar to the proof of Theorem 5.1. Similar to (5.14) we obtain:

$$
mnc_n + \lambda_m \sum_{i=1}^{n} \|\tilde{\tilde{\beta}}_i\|_1^{\alpha}
$$

$$
\leqslant m \sum_{i=1}^{n} \tilde{\tilde{\beta}}_i^{\mathsf{T}} \tilde{\Sigma}_i \tilde{\tilde{\beta}}_i + \lambda_m \sum_{i=1}^{n} \|\tilde{\tilde{\beta}}_i\|_1^{\alpha}
$$

$$
\leqslant m \sum_{i=1}^{n} \tilde{\tilde{\beta}}_i^{\mathsf{T}} \tilde{S}_i \tilde{\tilde{\beta}}_i + \lambda_m \sum_{i=1}^{n} \|\tilde{\tilde{\beta}}_i\|_1^{\alpha} + m\delta_m \sum_{i=1}^{n} \|\tilde{\tilde{\beta}}_i\|_1^2
$$

$$
\leqslant m \sum_{i=1}^{n} \tilde{\beta}_{i0}^{\mathsf{T}} \tilde{S}_i \tilde{\beta}_{i0} + \lambda_m \sum_{i=1}^{n} \|\tilde{\beta}_{i0}\|_1^{\alpha} + m\delta_m \sum_{i=1}^{n} \|\tilde{\tilde{\beta}}_i\|_1^2
$$

$$
\leqslant m \sum_{i=1}^{n} \tilde{\beta}_{i0}^{\mathsf{T}} \tilde{\Sigma}_i \tilde{\beta}_{i0} + \lambda_m \sum_{i=1}^{n} \|\tilde{\beta}_{i0}\|_1^{\alpha} + m\delta_m \sum_{i=1}^{n} \|\tilde{\beta}_{i0}\|_1^2 + m\delta_m \sum_{i=1}^{n} \|\tilde{\tilde{\beta}}_i\|_1^2
$$

$$
= mnc_n + \lambda_m \sum_{i=1}^{n} \|\tilde{\beta}_{i0}\|_1^{\alpha} + m\delta_m \sum_{i=1}^{n} \|\tilde{\beta}_{i0}\|_1^2 + m\delta_m \sum_{i=1}^{n} \|\tilde{\tilde{\beta}}_i\|_1^2.
$$

That is,

$$
\sum_{i=1}^{n} \left(\lambda_m \|\tilde{\tilde{\beta}}_i\|_1^{\alpha} - m\delta_m \|\tilde{\tilde{\beta}}_i\|_1^2 \right) \leqslant \lambda_m \sum_{i=1}^{n} \|\tilde{\beta}_{i0}\|_1^{\alpha} + m\delta_m \sum_{i=1}^{n} \|\tilde{\beta}_{i0}\|_1^2
$$
$$
= \mathcal{O}(mn\delta_m).
\tag{5.16}
$$

It is easy to see that the maximum of $\sum_{i=1}^{n} \|\tilde{\tilde{\beta}}_i\|_1^2$ subject to the constraint (5.16) is achieved when $\|\tilde{\tilde{\beta}}_1\|_1^2 = \cdots = \|\tilde{\tilde{\beta}}_n\|_1^2$. That is when $\|\tilde{\tilde{\beta}}_i\|_1^2$ solves $\lambda_m u^{\alpha} - m\delta_m u^2 = \mathcal{O}(m\delta_m)$. As $\lambda_n = \mathcal{O}(m\delta_m)$, the solution satisfies $u = \mathcal{O}(m\delta_m/\lambda_m)^{1/(\alpha-2)}$.

Hence we can conclude from (5.16)

$$\sum_{i=1}^{n} \|\tilde{\beta}_i\|_2^2 = \mathcal{O}(n(m\delta_m/\lambda_m)^{2/(\alpha-2)})$$

We now proceed similar to (5.15)

$$
\begin{aligned}
\sum_{i=1}^{n} \tilde{\beta}_i^{\mathsf{T}} \Sigma_i \tilde{\beta}_i &\leqslant \sum_{i=1}^{n} \tilde{\beta}_i^{\mathsf{T}} \tilde{S}_i \tilde{\beta}_i + \delta_m \sum_{i=1}^{n} \|\tilde{\beta}_i\|_1^2 \\
&\leqslant \sum_{i=1}^{n} \tilde{\beta}_{i0}^{\mathsf{T}} \tilde{S}_i \tilde{\beta}_{i0} + \frac{\lambda_m}{m} \sum_{i=1}^{n} \|\tilde{\beta}_{i0}\|_1^\alpha - \frac{\lambda_m}{m} \sum_{i=1}^{n} \|\tilde{\beta}_i\|_1^\alpha + \delta_m \sum_{i=1}^{n} \|\tilde{\beta}_i\|_1^2 \\
&\leqslant \sum_{i=1}^{n} \tilde{\beta}_{i0}^{\mathsf{T}} \tilde{\Sigma}_i \tilde{\beta}_{i0} + \frac{\lambda_m}{m} \sum_{i=1}^{n} \|\tilde{\beta}_{i0}\|_1^\alpha + \delta_m \sum_{i=1}^{n} \|\tilde{\beta}_{i0}\|_1^2 + \delta_m \sum_{i=1}^{n} \|\tilde{\beta}_i\|_1^2 \\
&\leqslant \sum_{i=1}^{n} \tilde{\beta}_{i0}^{\mathsf{T}} \tilde{\Sigma}_i \tilde{\beta}_{i0} + \mathcal{O}_p(n(m/\lambda_m)^{2/(\alpha-2)} \delta_m^{\alpha/(\alpha-2)}),
\end{aligned}
$$

since $\lambda_n = \mathcal{O}(m\delta_m)$.

Proof (Theorem 5.3). The proof follows that of Lemma 3.1 in Lounici et al. [7]. We start with (a) and (b). Since $\hat{\beta}$ minimizes (5.7), then, $\forall \beta$

$$\sum_{i=1}^{n} \|Y_i - X_i^{\mathsf{T}} \hat{\beta}_i\|_2^2 + \lambda \sum_{i=1}^{n} \|\hat{\beta}_i\|_1^\alpha \leqslant \sum_{i=1}^{n} \|Y_i - X_i^{\mathsf{T}} \beta_i\|_2^2 + \lambda \sum_{i=1}^{n} \|\beta_i\|_1^\alpha,$$

and hence, for $Y_i = X_i^{\mathsf{T}} \beta_i + \varepsilon_i$,

$$\sum_{i=1}^{n} \|X_i^{\mathsf{T}}(\hat{\beta}_i - \beta_i)\|_2^2 \leqslant \sum_{i=1}^{n} \left[2\varepsilon_i^{\mathsf{T}} X_i^{\mathsf{T}}(\beta_i - \hat{\beta}_i) + \lambda (\|\beta_i\|_1^\alpha - \|\hat{\beta}_i\|_1^\alpha) \right].$$

Denote $V_{i\ell} = \sum_{j=1}^{m} x_{ij\ell} \varepsilon_{ij} \sim \mathcal{N}(0, m\sigma^2)$, and introduce event $\mathscr{A}_i = \bigcap_{\ell=1}^{p} \{|V_{i\ell}| \leqslant \mu\}$, for some $\mu > 0$. Then

$$
\begin{aligned}
P(\mathscr{A}_i^c) &\leqslant \sum_{\ell=1}^{p} P(|V_{i\ell}| > \mu) \\
&= \sum_{\ell=1}^{p} 2 \left[1 - \Phi\{\mu/(\sigma\sqrt{m})\} \right] \\
&\leqslant p \exp\{-\mu^2/(2m\sigma^2)\}.
\end{aligned}
$$

For $\mathscr{A} = \bigcap_{i=1}^{n} \mathscr{A}_i$,

$$P(\mathscr{A}^c) = \sum_{i=1}^{n} P(\mathscr{A}_i^c) \leqslant pn \exp\{-\mu^2/(2m\sigma^2)\}.$$

Thus, if μ is large enough, $P(\mathscr{A}^c)$ is small, e.g., for $\mu = \sigma A (m \log(np))^{1/2}$, $A > \sqrt{2}$, we have $P(\mathscr{A}^c) \leqslant (np)^{1-A^2/2}$.

On event $\mathscr{A}$, for some $v > 0$,

$$\sum_{i=1}^{n} \left[||X_i^{\mathsf{T}}(\hat{\beta}_i - \beta_i)||_2^2 + v||\beta_i - \hat{\beta}_i||_1 \right]$$

$$\leqslant \sum_{i=1}^{n} \left[2\mu ||\beta_i - \hat{\beta}_i||_1 + \lambda (||\beta_i||_1^{\alpha} - ||\hat{\beta}_i||_1^{\alpha}) + v||\beta_i - \hat{\beta}_i||_1 \right]$$

$$= \sum_{i=1}^{n} \sum_{j=1}^{m} \left[\alpha \lambda \max(||\beta_i||_1^{\alpha-1}, ||\hat{\beta}_i||_1^{\alpha-1})(|\beta_{ij}| - |\hat{\beta}_{ij}|) + (v + 2\mu)|\beta_{ij} - \hat{\beta}_{ij}| \right]$$

$$\leqslant \sum_{i=1}^{n} \sum_{j=1}^{m} \left[\alpha \lambda \max(B^{\alpha-1}, \hat{B}^{\alpha-1})(|\beta_{ij}| - |\hat{\beta}_{ij}|) + (v + 2\mu)|\beta_{ij} - \hat{\beta}_{ij}| \right],$$

due to inequality $|x^{\alpha} - y^{\alpha}| \leqslant \alpha |x - y| \max(|x|^{\alpha-1}, |y|^{\alpha-1})$ which holds for $\alpha \geqslant 1$ and any x and y. To simplify the notation, denote $\mathscr{C} = \alpha \max(B^{\alpha-1}, \hat{B}^{\alpha-1})$.

Denote $J_i = J(\beta_i) = \{j : \beta_{ij} \neq 0\}$, $\mathscr{M}(\beta_i) = |J(\beta_i)|$. For each i and $j \in J(\beta_i)$, the expression in square brackets is bounded above by

$$[\lambda \mathscr{C} + v + 2\mu] |\beta_{ij} - \hat{\beta}_{ij}|,$$

and for $j \in J_i^c(\beta)$, the expression in square brackets is bounded above by 0, as long as $v + 2\mu \leqslant \lambda \mathscr{C}$:

$$-\lambda \mathscr{C} |\hat{\beta}_{ij}| + (v + 2\mu)|\hat{\beta}_{ij}| \leqslant 0.$$

This condition is satisfied if $v + 2\mu \leqslant \lambda \mathscr{C}$.

Hence, on $\mathscr{A}$, for $v + 2\mu \leqslant \lambda \mathscr{C}$,

$$\sum_{i=1}^{n} \left[||X_i^{\mathsf{T}}(\hat{\beta}_i - \beta_i)||_2^2 + v||\beta_i - \hat{\beta}_i||_1 \right] \leqslant \sum_{i=1}^{n} [\lambda \mathscr{C} + 2\mu + v] ||(\beta_i - \hat{\beta}_i)_{J_i}||_1.$$

This implies that

$$\sum_{i=1}^{n} ||X_i^{\mathsf{T}}(\hat{\beta}_i - \beta_i)||_2^2 \leqslant [\lambda \mathscr{C} + v + 2\mu] ||(\beta - \hat{\beta})_J||_1,$$

as well as that

$$||\beta - \hat{\beta}||_1 \leqslant \left[1 + \frac{2\mu}{v} + \frac{\lambda}{v} \mathscr{C} \right] ||(\beta - \hat{\beta})_J||_1.$$

Take $v = \lambda \mathscr{C}/2$, hence we need to assume that $2\mu \leqslant \lambda \mathscr{C}/2$:

$$\sum_{i=1}^{n} ||X_i^{\mathsf{T}}(\hat{\beta}_i - \beta_i)||_2^2 \leq \left[\frac{3\lambda}{2}\mathscr{C} + 2\mu\right] ||(\beta - \hat{\beta})_J||_1,$$

$$||\beta - \hat{\beta}||_1 \leq \left[3 + \frac{4\mu}{\lambda\mathscr{C}}\right] ||(\beta - \hat{\beta})_J||_1 \leq 4||(\beta - \hat{\beta})_J||_1.$$

(5.17)

which implies

$$||(\beta - \hat{\beta})_{J^c}||_1 \leq 3||(\beta - \hat{\beta})_J||_1.$$

Due to the generalized restricted eigenvalue assumption $\mathrm{RE}_1(s, 3, \kappa)$, $||X^{\mathsf{T}}(\beta - \hat{\beta})||_2 \geq \kappa\sqrt{m}||(\beta - \hat{\beta})_J||_2$, and hence, using (5.17),

$$||X^{\mathsf{T}}(\hat{\beta} - \beta)||_2^2 \leq \left[\frac{3\lambda}{2}\mathscr{C} + 2\mu\right] \sqrt{n\mathscr{M}(\beta)} ||(\hat{\beta} - \beta)_J||_2$$

$$\leq \left[\frac{3\lambda}{2}\mathscr{C} + 2\mu\right] \frac{\sqrt{n\mathscr{M}(\beta)}}{\kappa\sqrt{m}} ||X^{\mathsf{T}}(\hat{\beta} - \beta)||_2,$$

where $\mathscr{M}(\beta) = \max_i \mathscr{M}(\beta_i)$, implying that

$$||X^{\mathsf{T}}(\hat{\beta} - \beta)||_2 \leq \left[\frac{3\lambda}{2}\mathscr{C} + 2\mu\right] \frac{\sqrt{n\mathscr{M}(\beta)}}{\kappa\sqrt{m}}$$

$$= \frac{\sqrt{n\mathscr{M}(\beta)}}{\kappa\sqrt{m}} \left[\frac{3\lambda}{2}\mathscr{C} + 2A\sigma\sqrt{m\log(np)}\right].$$

Also,

$$||\beta - \hat{\beta}||_1 \leq 4||(\beta - \hat{\beta})_J||_1 \leq 4\frac{\sqrt{n\mathscr{M}(\beta)}}{\sqrt{m}\kappa} ||X^{\mathsf{T}}(\beta - \hat{\beta})||_2$$

$$\leq \frac{4n\mathscr{M}(\beta)}{m\kappa^2} \left[\frac{3\lambda}{2}\mathscr{C} + 2A\sigma\sqrt{m\log(np)}\right].$$

Hence, a) and b) of the theorem are proved.

(c) For i, ℓ: $\hat{\beta}_{i\ell} \neq 0$, we have

$$2X_{i\cdot\ell}(Y_i - X_i^{\mathsf{T}}\hat{\beta}_i) = \lambda\alpha\mathrm{sgn}\,(\hat{\beta}_{i\ell})||\hat{\beta}_i||_1^{\alpha-1},$$

where we used the notation that $X_{i\cdot\ell} = (X_{i1\ell}, \ldots, X_{im\ell})^{\mathsf{T}}$. Using inequality $|x - y| \geq ||x| - |y||$ easily derived from the triangle inequality, we have

$$\sum_{\ell:\hat{\beta}_{i\ell}\neq 0}\left[\sum_{j=1}^{m}\sum_{r=1}^{p}X_{ij\ell}X_{ijr}(\beta_{ir}-\hat{\beta}_{ir})\right]^{2} = \sum_{\ell:\hat{\beta}_{i\ell}\neq 0}\left[X_{i\cdot\ell}X_i^{\mathsf{T}}(\beta_i-\hat{\beta}_i)\right]^{2}$$

$$\geqslant \sum_{\ell:\hat{\beta}_{i\ell}\neq 0}\left(|X_{i\cdot\ell}(Y_i-X_i^{\mathsf{T}}\hat{\beta}_i)|-|X_{i\cdot\ell}(Y_i-X_i^{\mathsf{T}}\beta_i)|\right)^{2}$$

$$\geqslant \sum_{\ell:\hat{\beta}_{i\ell}\neq 0}\left(\alpha\lambda\|\hat{\beta}_i\|_1^{\alpha-1}/2-\mu\right)^{2}$$

$$= \mathscr{M}(\hat{\beta}_i)(\alpha\lambda\|\hat{\beta}_i\|_1^{\alpha-1}/2-\mu)^{2}.$$

Thus,

$$\mathscr{M}(\hat{\beta}_i) \leqslant \|X_i^{\mathsf{T}}(\beta_i-\hat{\beta}_i)\|_2^2\frac{m\phi_{i,\max}}{\left(\lambda\alpha\|\hat{\beta}_i\|_1^{\alpha-1}/2-\mu\right)^{2}}.$$

Theorem is proved.

Proof (Theorem 5.4.). To satisfy the conditions of Theorem 5.3, we can take $B=b$ and $\lambda = \frac{4A\sigma}{\alpha b^{\alpha-1}}\sqrt{m\log(np)}$. By Lemma A.1 in Bochkina & Ritov ([2]),

$$\frac{\lambda}{m\delta_n} = \frac{4A\sigma}{\alpha b^{\alpha-1}}\frac{1}{C\sigma^2}\sqrt{\frac{\log(np)}{m}}\sqrt{\frac{m\eta}{\log(n(p+1)^2)}} = C\frac{\sqrt{\eta}}{\alpha b^{\alpha-1}} \leqslant C_1,$$

hence assumption $\lambda = \mathcal{O}(m\delta_n)$ of Theorem 5.2 is satisfied.

¿From the proof of Theorem 5.2, it follows that

$$\|\hat{\beta}_i\|_1 = \mathcal{O}\left((m\delta_n/\lambda_n)^{1/(\alpha-2)}\right) = \mathcal{O}\left(\left(\frac{b^{\alpha-1}}{\sqrt{\eta}}\right)^{1/(\alpha-2)}\right).$$

Hence, we can take $B=b$ and $\hat{B} = C\left(\frac{b^{\alpha-1}}{\sqrt{\eta}}\right)^{1/(\alpha-2)}$ for some $C>0$, and apply Theorem 5.3. Then $\max(1,\hat{B}/B)$ is bounded by

$$\max\left[1,C\frac{b^{(\alpha-1)/(\alpha-2)-1}}{\eta^{1/(2(\alpha-2))}}\right] = \max\left[1,C\frac{b^{1/(\alpha-2)}}{\eta^{1/(2(\alpha-2))}}\right] = \left(\frac{Cb}{\sqrt{\eta}}\right)^{1/(\alpha-2)},$$

since $\frac{Cb}{\sqrt{\eta}} \geqslant C_2\frac{\eta^{1/(2(\alpha-1))}}{\sqrt{\eta}} \geqslant C_2\eta^{-(\alpha-2)/(2(\alpha-1))}$ is large for small η.

Thus,

$$\frac{3\alpha\lambda}{2\sqrt{m}}\max(B^{\alpha-1},\hat{B}^{\alpha-1})+2A\sigma\sqrt{\log(np)}$$

$$\leqslant 6AC\sigma\sqrt{\log(np)}\frac{b^{(\alpha-1)/(\alpha-2)}}{\eta^{(\alpha-1)/(2(\alpha-2))}}+2A\sigma\sqrt{\log(np)}$$

$$=2A\sigma\sqrt{\log(np)}\left[3C\left(\frac{b}{\sqrt{\eta}}\right)^{(\alpha-1)/(\alpha-2)}+1\right],$$

and, applying Theorem 5.3, we obtain (a) and (b).

c) Apply c) in Theorem 5.3, summing over $i \in \mathscr{I}$:

$$\sum_{i\in\mathscr{I}}\mathscr{M}(\hat{\beta}_i)\leqslant \|X^{\mathsf{T}}(\beta-\hat{\beta})\|_2^2\frac{m\phi_{\max}}{(\mu\delta)^2}$$

$$\leqslant \frac{4sn\phi_{\max}}{\kappa^2\delta^2}\left[1+3C\left(\frac{b}{\sqrt{\eta}}\right)^{(\alpha-1)/(\alpha-2)}\right]^2.$$

References

1. Bickel, P., Ritov, Y., Tsybakov, A.: Simultaneous analysis of Lasso and Dantzig selector. Ann. Statist. **37**, 1705–1732 (2009)
2. Bochkina, N., Ritov, Y.: Sparse empirical Bayes analysis (2009). URL http://arxiv. org/abs/0911.5482
3. Brown, L., Greenshtein, E.: Nonparametric empirical Bayes and compound decision approaches to estimation of a high-dimensional vector of normal means. Ann. Statist. **37**, 1685–1704 (2009)
4. Greenshtein, E., Park, J., Ritov, Y.: Estimating the mean of high valued observations in high dimensions. J. Stat. Theory Pract. **2**, 407–418 (2008)
5. Greenshtein, E., Ritov, Y.: Persistency in high dimensional linear predictor-selection and the virtue of over-parametrization. Bernoulli **10**, 971–988 (2004)
6. Greenshtein, E., Ritov, Y.: Asymptotic efficiency of simple decisions for the compound decision problem. In: J. Rojo (ed.) Optimality: The 3rd Lehmann Symposium, IMS Lecture-Notes Monograph series, vol. 1, pp. 266–275 (2009)
7. Lounici, K., Pontil, M., Tsybakov, A.B., van de Geer, S.: Taking advantage of sparsity in multi-task learning. In: Proceedings of COLT'09, pp. 73–82 (2009)
8. Robbins, H.: Asymptotically subminimax solutions of compound decision problems. In: Proceedings of the 2nd Berkeley Symposium on Mathematical Statistics and Probability, vol. 1, pp. 131–148 (1951)
9. Robbins, H.: An empirical Bayes approach to statistics. In: Proceedings of the 3rd Berkeley Symposium on Mathematical Statistics and Probability, vol. 1, pp. 157–163 (1956)
10. Tibshirani, R.: Regression shrinkage and selection via the Lasso. J. R. Stat. Soc. Ser. B Stat. Methodol. **58**, 267–288 (1996)
11. Yuan, M., Lin, Y.: Model selection and estimation in regression with grouped variables. J. R. Stat. Soc. Ser. B Stat. Methodol. **68**, 49–67 (2006)
12. Zhang, C.H.: Compound decision theory and empirical Bayes methods. Ann. Statist. **31**, 379–390 (2003)
13. Zhang, C.H.: General empirical Bayes wavelet methods and exactly adaptive minimax estimation. Ann. Statist. **33**, 54–100 (2005)

Part V
Invited and Contributed Talks Given During the Summer School

Appendix A
List of the Courses

Laurent Cavalier (Université Aix-Marseille I)
Inverse Problems in Statistics

Victor Chernozhukov (Massachussets Institute of Technology)
High Dimensional Statistical Estimation with Applications to Economics

Appendix B
List of the Invited Talks

Felix Abramovich   (Tel Aviv University)
Bayesian Multiple Testing and Testimation in High-Dimensional Settings

Xiaohong Chen   (Yale University)
On Plug-In Estimation of Functionals of Semi/Nonparametric Conditional and Unconditional Moment Models

Rama Cont   (CNRS – Columbia University)
Solving Ill-Posed Inverse Problems Using Minimal Entropy Random Mixtures: Application to Inverse Problems in Financial Modeling

Jean-Pierre Florens   (Université Toulouse I)
Nonparametric Instrumental Variables

Emmanuel Guerre   (Queen Mary, University of London)
Semiparametric Estimation of First-Price Auctions with Risk Averse Bidders

Joel Horowitz   (Northwestern University)
Confidence Bands for Functions Estimated by Nonparametric Instrumental Variables

Yuichi Kitamura   (Yale University)
Nonparametric Methods for Econometric Models with Unobserved Heterogeneity

Jean-Michel Loubes   (Université Paul Sabatier, Toulouse)
Tests for Inverse Problems

Ya'acov Ritov   (The Hebrew University of Jerusalem)
Future Observations, Present Means, and Past Parameters: Complex Models and Statistical Inference

Jean-Marc Robin   (Université Paris Panthéon-Sorbonne – University College London)
Semi-parametric Estimation of (Noisy) Independent Factor Models

Appendix C
List of the Contributed Talks and Posters

Contributed Talks

Brendan Beare   (University of California, San Diego)
Distributional Replication

Christine De Mol   (Université Libre de Bruxelles)
Sparse and Stable Markowitz Portfolios

Kirill Evdokimov   (Yale University)
Identification and Estimation of a Nonparametric Panel Data Model with Unobserved Heterogeneity

Jan Johannes   (University of Heidelberg)
On Rate Optimal Local Estimation in Nonparametric Instrumental Regression

Karim Lounici   (ENSAE-CREST – Université Paris Diderot)
Taking Advantage of Sparsity in Multi-task Learning

Clément Marteau   (INSA Toulouse)
Oracle Inequality for Instrumental Variable Regression

Arnaud Maurel   (ENSAE-CREST)
Inference on a Generalized Roy Model with Exclusion Restrictions

Maria-Augusta Miceli   (Università di Roma La Sapienza)
Large Dimension Forecasting Models and Random Singular Value Spectra

Anna Simoni   (Toulouse School of Economics)
On the Regularization Power of the Prior Distribution in Linear Ill-Posed Inverse Problems

Victoria Zinde-Walsh   (McGill University)
Errors-in-Variables Models: A Generalized Functions Approach

Contributed Posters

Dirk Antonczyk   (Universität Freiburg)
A Nonparametric Additive Model Identifying Age, Time, and Cohort Effects

Nicolas Brunel   (ENSIIE and Université d'Evry)
Parameter Estimation in Ordinary Differential Equations with Orthogonality Conditions

Guillaume Chevillon   (ESSEC and ENSAE-CREST)
Learning to Generate Long Memory

Willem Kruijer   (Université Paris-Dauphine and ENSAE-CREST)
Adaptive Bayesian Density Estimation with Location Scale Mixtures

Vitaliy Oryshchenko   (Cambridge University)
Density Forecasts and Shrinkage

Maria Putintseva   (Universität Zürich)
What Could We Infer from Prediction Market Prices?

Subramanian Ramamoorthy   (University of Edinburgh)
An Online Algorithm for Multi-strategy Trading Utilizing Market Regimes

Alex Stuckey   (University of Western Australia)
A Single-Index Model for Spatial Data